Robert Lawlor

Sacred Geometry

Philosophy & Practice

with 202 illustrations and diagrams,
56 in two colours

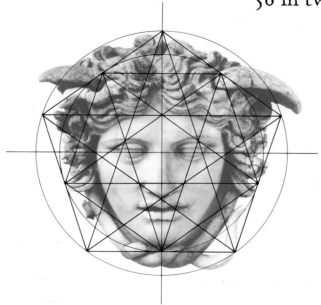

T&H

For R. A. Schwaller de Lubicz and Lucie Lamy

This book originated in a series of seminars held in New York City for the Lindisfarne Association, Crestone, Colorado.

Diagrams by Melvyn Bernstein, A.I.A.
Illustration on p.1: see p.53.

ART AND IMAGINATION

First published in the United Kingdom in 1982
by Thames & Hudson Ltd, 181A High Holborn,
London WC1V 7QX

This edition first published in the United States of America
in 1989 by Thames & Hudson Inc., 500 Fifth Avenue,
New York, New York 10110

Reprinted 2021

Sacred Geometry © 1982 Thames & Hudson Ltd, London

British Library Cataloguing-in-Publication Data
A catalogue record for this book is available from the
British Library

Library of Congress Control Number 88-51328

ISBN 978-0-500-81030-9

Printed and bound in China by Everbest Printing Co. Ltd

Contents

Introduction

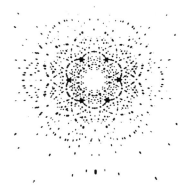

X-ray diffraction pattern in beryl, indicating a patterned array of intervals surrounding a central node much like the pattern of partial overtones around a fundamental tone.

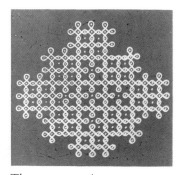

These geometric array patterns called *kolams* are drawn, with powdered chalk, by South Indian women on the doorstep each morning, to evoke the spirit of order and harmony into the home.

In science today we are witnessing a general shift away from the assumption that the fundamental nature of matter can be considered from the point of view of substance (particles, quanta) to the concept that the fundamental nature of the material world is knowable only through its underlying patterns of wave forms.

Both our organs of perception and the phenomenal world we perceive seem to be best understood as systems of pure pattern, or as geometric structures of form and proportion. Therefore, when many ancient cultures chose to examine reality through the metaphors of geometry and music (music being the study of the proportional laws of sound frequency), they were already very close to the position of our most contemporary science.

Professor Amstutz of the Mineralogical Institute at the University of Heidelberg recently said:

> Matter's latticed waves are spaced at intervals corresponding to the frets on a harp or guitar with analogous sequences of overtones arising from each fundamental. The science of musical harmony is in these terms practically identical with the science of symmetry in crystals.

The point of view of modern force-field theory and wave mechanics corresponds to the ancient geometric-harmonic vision of universal order as being an interwoven configuration of wave patterns. Bertrand Russell, who began to see the profound value of the musical and geometric base to what we now call Pythagorean mathematics and number theory, also supported this view in *The Analysis of Matter*: 'What we perceive as various qualities of matter,' he said, 'are actually differences in periodicity.'

In biology, the fundamental role of geometry and proportion becomes even more evident when we consider that moment by moment, year by year, aeon by aeon, every atom of every molecule of both living and inorganic substance is being changed and replaced. Every one of us within the next five to seven years will have a completely new body, down to the very last atom. Amid this constancy of change, where can we find the basis for all that which appears to be consistent and stable? Biologically we may look to our ideas of genetic coding as the vehicle of replication and continuity, but this coding does not lie in the particular atoms (or carbon, hydrogen, oxygen and nitrogen) of which the gene substance, DNA, is composed; these are all also subject to continual change and replacement. Thus the carrier of continuity is not only the molecular composition of the DNA, but also its helix form. This form is responsible for the replicating power of the DNA. The helix, which is a special type from the group of regular spirals, results from sets of fixed geometric proportions, as we shall see in detail later on. These proportions can be understood to exist *a priori*, without any material counterpart, as abstract, geometric relationships. The architecture of bodily existence is determined by an invisible, immaterial world of pure form and geometry.

Modern biology increasingly recognizes the importance of the form and the bonding relationships of the few substances which comprise the molecular body of living organisms. Plants, for example, can carry out the process of photosynthesis only because the carbon, hydrogen, nitrogen and magnesium of the chlorophyll molecule are arranged in a complex twelvefold symmetrical pattern, rather like that of a daisy. It seems that the same constituents in any other arrangement cannot transform the radiant energy of light into life substance. In mythological thought, twelve most often occurs as the number of the universal mother of life, and so this twelvefold symbol is precise even to the molecular level.

The specialization of cells in the body's tissue is determined in part by the spatial position of each cell in relation to other cells in its region, as well as by an informational image of the totality to which it belongs. This spatial awareness on a cellular level may be thought of as the innate geometry of life.

All our sense organs function in response to the geometrical or proportional – not quantitative – differences inherent in the stimuli they receive. For example, when we smell a rose we are not responding to the chemical substances of its perfume, but instead to the geometry of their molecular construction. That is to say, any chemical substance that is bonded together in the same geometry as that of the rose will smell as sweet. Similarly, we do not hear simple quantitative differences in sound wave frequencies, but rather the logarithmic, proportional differences between frequencies, logarithmic expansion being the basis of the geometry of spirals.

Our visual sense differs from our sense of touch only because the nerves of the retina are not tuned to the same range of frequencies as are the nerves embedded in our skin. If our tactile or haptic sensibilities were responsive to the same frequencies as our eyes, then all material objects would be perceived to be as ethereal as projections of light and shadow. Our different perceptual faculties such as sight, hearing, touch and smell are a result then of various proportioned reductions of one vast spectrum of vibratory frequencies. We can understand these proportional relationships as a sort of geometry of perception.

With our bodily organization into five or more separate perceptual thresholds, there is seemingly little in common between visual space, auditory space and haptic space, and there seems to be even less connection between these physiological spaces and pure, abstract metric or geometric space, not to mention here the differing awareness of phychological space. Yet all these modes of spatial being converge in the human mind-body. Within the human consciousness is the unique ability to perceive the transparency between absolute, permanent relationships, contained in the insubstantial forms of a geometric order, and the transitory, changing forms of our actual world. The content of our experience results from an immaterial, abstract, geometric architecture which is composed of harmonic waves of energy, nodes of relationality, melodic forms springing forth from the eternal realm of geometric proportion.

From the apparent world to the subatomic, all forms are only envelopes for geometric patterns, intervals and relationships.

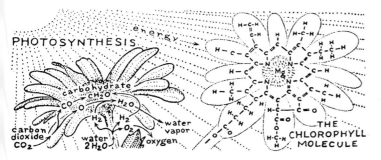

Here we find twelvefold symmetry as the life-giver or womb which transforms light into the basic spectrum of organic substance. This is recalled symbolically in the stained-glass window, which transforms light into the colour spectrum.

I The Practice of Geometry

'What is God? He is length, width, height and depth.'
ST BERNARD OF CLAIRVAUX, *On Consideration*

'Geometry' means 'measure of the earth'. In ancient Egypt, from which Greece inherited this study, the Nile would flood its banks each year, covering the land and obliterating the orderly marking of plot and farm areas. This yearly flood symbolized to the Egyptian the cyclic return of the primal watery chaos, and when the waters receded the work began of redefining and re-establishing the boundaries. This work was called geometry and was seen as a re-establishment of the principle of order and law on earth. Each year the areas measured out would be somewhat different. The human order would shift and this was reflected in the ordering of the earth. The Temple astronomer might say that certain celestial configurations had changed so that the orientation or location of a temple had to be adjusted accordingly. So the laying of squares upon the earth had, for the Egyptian, a metaphysical as well as a physical and social dimension. This activity of 'measuring the earth' became the basis for a science of natural law as it is embodied in the archetypal forms of circle, square and triangle.

Geometry is the study of *spatial order* through the measure and relationships of forms. Geometry and arithmetic, together with astronomy, the science of *temporal order* through the observation of cyclic movement, constituted the major intellectual disciplines of classical education. The fourth element of this great fourfold syllabus, the Quadrivium, was the study of harmony and music. The laws of simple harmonics were considered to be universals which defined the relationship and interchange between the temporal movements and events of the heavens and the spatial order and development on earth.

The implicit goal of this education was to enable the mind to become a channel through which the 'earth' (the level of manifested form) could receive the abstract, cosmic life of the heavens. The practice of geometry was an approach to the way in which the universe is ordered and sustained. Geometric diagrams can be contemplated as still moments revealing a continuous, timeless, universal action generally hidden from our sensory perception. Thus a seemingly common mathematical activity can become a discipline for intellectual and spiritual insight.

Plato considered geometry and number as the most reduced and essential, and therefore the ideal, philosophical language. But it is only by virtue of functioning at a certain 'level' of reality that geometry and number can become a vehicle for philosophic contemplation. Greek philosophy defined this notion of 'levels', so useful in our thinking, distinguishing the 'typal' and the 'archetypal'. Following the indication given by the Egyptian wall reliefs, which are laid out in three registers, an upper, a middle and a lower, we can define a third level, the ectypal, situated between the archetypal and the typal.

To see how these operate, let us take an example of a tangible thing, such as the bridle of a horse. This bridle can have a number of forms, materials, sizes, colours, uses, all of which are bridles. The bridle considered in this way, is typal; it is existing, diverse and variable. But on another level there is the idea or form of the bridle, the guiding model of all bridles. This is an unmanifest, pure, formal idea and its level is ectypal. But yet above this there is the archetypal level which is that of the *principle* or *power-activity*. that is a *process* which the ectypal form and typal example of the bridle only represent. The archetypal is concerned with universal processes or dynamic patterns which can be considered independently of any structure or

Geometry as a contemplative practice is personified by an elegant and refined woman, for geometry functions as an intuitive, synthesizing, creative yet exact activity of mind associated with the feminine principle. But when these geometric laws come to be applied in the technology of daily life they are represented by the rational, masculine principle: contemplative geometry is transformed into practical geometry.

ABOVE *Arithmetic* is also personified as a woman, but not as grand and noble in attire as *Geometry*, perhaps symbolically indicating that Geometry was considered as a higher order of knowledge. On her thighs (symbolizing the generative function) are two geometric progressions. The first series, 1, 2, 4, 8, goes down the left thigh, associating the even numbers with the feminine, passive side of the body. The second series, 1, 3, 9, 27, goes down the right thigh, associating the odd numbers with the masculine, active side, an association which goes back to the Pythagoreans, who called the odd numbers male and the even female. The Greeks called these two series the *Lambda*, and Plato in the *Timaeus* uses them to describe the World Soul (see p. 83). On the woman's left sits Pythagoras using an abacus system for computation. In this system, number notation is still dependent upon spatial arrangement. Boethius sits on her right using Arabic numerals in a modern system of calculation in which number notation has become a separate, abstract system independent of its geometric origin.

BELOW Pythagoras is credited with first establishing the relationship between number ratios and sound frequencies. He is shown here experimenting with bells, water-glasses, stretched cords, and various sized pipes; his Hebrew counterpart, Jubal, uses weighted hammers on an anvil. The whole number ratios for determining the consonant sounds in a musical scale are either drawn from or are multiples of the numbers in the two progressions of the *Lambda*.

The ancient astronomers designated the movement and position of celestial bodies through angular notation. The varied angular positions of the sun, moon, planets and stars were related to the cyclic changes in the natural world, such as moon phases, seasons, tides, plant growth, human and animal fertility, etc. It was the angle which specified the influences of celestial patterns on earthly events. (In this way we can appreciate the similar root of the words *angle* and *angel*.) Today the newly emerging science of heliobiology verifies that the angular position of the moon and planets does affect the electromagnetic and cosmic radiations which impact with the earth, and in turn these field fluctuations affect many biological processes.

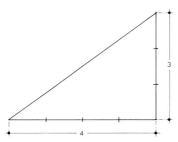

In ancient trigonometry an angle is a relationship between two whole numbers. In this example the angle at left is an expression of the ratio 3 to 4, and with this system spatial coordinates can easily be put into relationship with sound frequencies, such as the musical fourth (see p. 85).

material form. Modern thought has difficult access to the concept of the archetypal because European languages require that verbs or action words be associated with nouns. We therefore have no linguistic forms with which to image a process or activity that has no material carrier. Ancient cultures symbolized these pure, eternal processes as gods, that is, powers or lines of action through which Spirit is concretized into energy and matter. The bridle, then, relates to archetypal activity through the function of *leverage*; the principle that *energies are controlled, specified and modified through the effects of angulation*.

Thus we find that often the angle – which is fundamentally a relationship of two numbers – would have been used in ancient symbolism to designate a group of fixed relationships controlling interacting complexes or patterns. Thus the archetypes or gods represent dynamic functions forming links between the higher worlds of constant interaction and process and the actual world of particularized objects. We find, for example, that a 60° angle has quite different structural and energetic properties from an angle of 90° or of 45°. Likewise, geometric optics reveals that each substance characteristically refracts light at its own particular angle, and it is this angle which gives us our most precise definition of the substance. Furthermore, the angles in the bonding patterns of molecules determine to a great extent the qualities of the substance.

In the case of the bridle, this angulation or angular play is manifested in the relation of the bit to the bridle strap, or between the bit and the bend of the horse's neck and jaw, both controlled by the angulation between the forearm and the biceps of the rider. From the level of the archetype or active Idea, the principle of the bridle can be applied metaphorically to many regions of human experience. For instance, when St Paul describes the process of self-discipline by which a higher intentionality attempts to control the lower, 'animal' nature, he says that when one can bridle the mouth he can then master the rest of his nature. But while at the archetypal level this image can be metaphysically and poetically expansive, it also finds its exact, geometrical representation in the *angle*. It is the precise angle of the arm in play with the angle of the bridle that controls the energy of the horse.

Functioning then at the archetypal level, Geometry and Number describe fundamental, causal energies in their interwoven, eternal dance. It is this way of seeing that stands behind the expression of cosmological systems as geometric configurations. For example, the most revered of all Tantric diagrams, the Sri Yantra, images all the necessary functions active in the universe through its nine interlocked triangles. To immerse oneself in such a geometric diagram is to enter into a kind of philosophic contemplation.

For Plato, Reality consisted of pure essences or archetypal Ideas, of which the phenomena we perceive are only pale reflections. (The Greek work 'Idea' is also translated as 'Form'.) These Ideas cannot be perceived by the senses, but by pure reason alone. Geometry was the language recommended by Plato as the clearest model by which to describe this metaphysical realm.

> And do you not know that they [the geometers] make use of the visible forms and talk about them, though they are not of them but of those things of which they are a likeness, pursuing their inquiry for the sake of the square as such and the diagonal as such, and not for the sake of the image of it which they draw? And so on in all cases . . . What they really seek is to get sight of those realities which can be seen only by the mind.   PLATO, *Republic*, VII, 510 d, e.

The Platonist sees our geometrical knowledge as innate in us, having been acquired before birth when our souls were in contact with the realm of ideal being.

> All mathematical forms have a primary subsistence in the soul; so that prior to the sensible she contains self-motive numbers; vital figures prior to such as are apparent; harmonic ratios prior to things harmonized; and invisible circles prior to the bodies that are moved in a circle.   THOMAS TAYLOR

Plato demonstrates this in the *Meno* where he has an untutored servant boy solve by intuition the geometric problem of the doubling of the square.

The Sri Yantra is drawn from nine triangles, four pointed downward and five pointed upward, thus forming 42 (6 × 7) triangular fragments around a central triangle. There is probably no other set of triangles which interlock with such integrational perfection.

For the human spirit caught within a spinning universe in an ever confusing flow of events, circumstance and inner turmoil, to seek truth has always been to seek the invariable, whether it is called Ideas, Forms, Archetypes, Numbers or Gods. To enter a temple constructed wholly of invariable geometric proportions is to enter an abode of eternal truth. Thomas Taylor says, 'Geometry enables its votary, like a bridge, to pass over the obscurity of material nature, as over some dark sea to the luminous regions of perfect reality.' Yet this is by no means an automatic happening that occurs just by picking up a geometry book. As Plato says, the soul's fire must gradually be rekindled by the effort:

> You amuse me, you who seem worried that I impose impractical studies upon you. It does not only reside with mediocre minds, but all men have difficulty in persuading themselves that it is through these studies, as if with instruments, that one purifies the eye of the soul, and that one causes a new fire to burn in this organ which was obscured and as though extinguished by the shadows of the other sciences, an organ whose conservation is more important than ten thousand eyes, since it is by it alone that we contemplate the truth.
>
> *Republic*, VII, 527 d, e
> (as quoted by Theon of Smyrna (2nd c. AD) in his
> *Mathematics Useful for Understanding Plato*)

Geometry deals with pure form, and philosophical geometry re-enacts the unfolding of each form out of a preceding one. It is a way by which the essential creative mystery is rendered visible. The passage from creation to procreation, from the unmanifest, pure, formal idea to the 'here-below', the world that spins out from that original divine stroke, can be mapped out by geometry, and experienced through the practice of geometry: this is the purpose of the 'Workbook' sections of this book.

Inseparable from this process is the concept of Number, and, as we shall see, for the Pythagorean, Number and Form at the ideal level were one. But number in this context must be understood in a special way. When Pythagoras said, 'All is arranged according to Number', he was not thinking of numbers in the ordinary, enumerative sense. In addition to simple *quantity*, numbers on the ideal level are possessed of *quality*, so that 'twoness', 'threeness' or 'fourness', for example, are not merely composed of 2, 3, or 4 units, but are wholes or unities in themselves, each having related powers. 'Two', for instance, is seen as the original essence from which the *power of duality* proceeds and derives its reality.

R.A. Schwaller de Lubicz gives an analogy by which this universal and archetypal sense of Number can be understood. A revolving sphere presents us with the notion of an axis. We think of this axis as an ideal or imaginary line through the sphere. It

The twelfth-century architecture of the Cistercian Order achieves its visual beauty through designs which conform to the proportional system of musical harmony. Many of the abbey churches of this period were acoustic resonators transforming a human choir into celestial music. St Bernard of Clairvaux, who inspired this architecture, said of their design, 'There must be no decoration, only proportion.'

Christ is shown using compasses to re-enact the creation of the universe from the chaos of the primal state. This icon can also be understood as an image of individual self-creation; for here, as in many medieval images of Christ, Tantric symbolism is evident. Christ holds the compass with his hand across the vital centre called the heart chakra, and from this centre he organizes the turmoil of the vital energies contained in the lower chakras which are indicated on the body by centres at the navel and genitals. Geometry is symbolized here in both the individual and universal sense as an instrument through which the higher archetypal realm transmits order and harmony to the vital and energetic worlds.

has no objective existence, yet we cannot help but be convinced of its reality; and to determine anything about the sphere, such as its inclination or its speed of rotation we must refer to this imaginary axis. Number in the enumerative sense corresponds to the measures and movements of the outer surface of the sphere, while the universal aspect of Number is analogous to the immobile, unmanifest, functional principle of its axis.

Let us shift our analogy to the two-dimensional plane. If we take a circle and a square and give the value 1 to the diameter of the circle and also to the side of the square, then the diagonal of the square will always be (and this is an invariable law) an 'incommensurable', 'irrational' number. It is said that such a number can be carried out to an infinite number of decimal places without ever arriving at a resolution. In the case of the diagonal of the square, this decimal is 1·4142 . . . and is called the square root of 2, or $\sqrt{2}$. With the circle, if we give the diameter the value 1, the circumference will also always be of the incommensurable type, 3·14159 . . . which we know by the Greek symbol π, *pi*.

The principle remains the same in the inversion: if we give the fixed, rational value of 1 to the diagonal of the square and to the circumference of the circle, then the side of the square and the radius of the circle will become of the incommensurable 'irrational' type: $1/\sqrt{2}$ and $1/\pi$.

It is exactly at this point that quantified mathematics and geometry go their separate ways, because numerically we can never know exactly the diagonal of the square nor the circumference of the circle. Yes, we can round-off after a certain number of decimal places, and treat these cut off numbers like any other number, but we can never reduce them to a quantity. In geometry, however, the diagonal and the circumference, when considered in the context of *formal relationship* (diagonal to side; circumference to diameter), are absolutely knowable, self-evident realities: $1:\sqrt{2}$ and $1:\pi$. Number is considered as a *formal relationship*, and this type of numerical relationship is called a *function*. The square root of 2 is the functional number of a square. *Pi* is the functional number of a circle. Philosophic geometry – and consequently sacred art and architecture – is very much concerned with these 'irrational' functions, for the simple reason that they demonstrate graphically a level of experience which is universal and invariable.

The irrational functions (which we will consider rather as supra-rational) are a key opening a door to a higher reality of Number. They demonstrate that Number is above all a relationship; and no matter what quantities are applied to the side and to the diameter the relationship will remain invariable, for in essence this functional aspect of Number is neither large nor small, neither infinite nor finite: it is universal. Thus within the concept of Number there is a definite, finite, particularizing power and also a universal synthesizing power. One may be called the exoteric or external aspect of number, the other the esoteric or inner, functional aspect.

Let us look at the first four primary numbers in this spirit.

The number ONE can of course define a quantity; as, for example, one apple. But in its other sense, it perfectly represents the principle of absolute unity, and as such has often been used as the symbol to represent God. As a statement of form it can in one sense represent a point – it has been called the 'pointal' number, the *bindu* or seed in the Hindu mandala – or in another sense it can represent the perfect circle.

TWO is a quantity, but symbolically it represents, as we have already seen, the principle of Duality, the power of multiplicity. At the same time it has its formal sense in the representation of a line, in that two points define a line.

THREE is a quantity, but as a principle it represents the Trinity, a vital concept which we will meet again later. Its formal sense is that of the triangle, which is formed from three points. With three a qualitative transition is made from the pure, abstract elements of point and line to the tangible, measurable state which is called a *surface*. In India the triangle was called the Mother, for it is the membrane or birth channel through which all the transcendent powers of unity and its initial division into polarity must pass in order to enter into the manifest realm of surface. The triangle acts as the mother of form.

But three is yet only a principle of creation, forming the passage between the transcendent and the manifest realms, whereas FOUR represents at last the 'first born thing', the world of Nature, because it is the product of the procreative process, that is of multiplication: $2 \times 2 = 4$. As a form, four is the square, and represents materialization.

The universality of Number can be seen in another, more physical context. We learn from modern physics that from gravity to electromagnetism, light, heat, and even in what we think of as solid matter itself, the entire perceptible universe is composed of vibrations, perceived by us as wave phenomena. Waves are pure temporal patterns, that is dynamic configurations composed of amplitude, interval and frequency, and they can be defined and understood by us only through Number. Thus our whole universe is reducible to Number. Every living body physically

This Japanese Zen calligraphic drawing beautifully shows 'creation' through the simple progression from the Unity of the circle, through the triangle, to the manifest form of the square.

vibrates, all elemental or inanimate matter vibrates molecularly or atomically, and every vibrating body emits a sound. The study of sound, as the ancients intuited, provides a key to the understanding of the universe.

We've noted already that the ancients gave considerable attention to the study of musical harmony in relation with the study of mathematics and geometry. The origin of this tradition is generally associated with Pythagoras (560–490 BC) and his school, yet Pythagoras may be considered as a window through which we can glimpse the quality of the intellectual world of an older, eastern and mideastern tradition. For this line of thinking, the sounding of the octave (an octave is for example two successive 'Do's' on a musical scale) was the most significant moment of all contemplation. It represented the beginning and goal of creation. What happens when we sound the perfect octave? There is an immediate, simultaneous coinciding of understanding which has occurred on several levels of being. Without any intervention of thought or concept or image, we immediately recognize the recurrence of the initial tone in the form of the octave. It is the same note, yet it is different; it is the completion of a cycle, a spiral from seed to new seed. This timeless, instantaneous recognition (more accurate than any visual recognition) is universal among humans.

But something else has happened as well. A guitarist sounds a string. He next depresses this string with his finger exactly at its midpoint. He sounds the half-string. The frequency of vibrations produced is double that given by the whole string, and the tone is raised by one octave. The string length has been divided by two, and the number of vibrations per second has been multiplied by two: 1/2 has created its mirror opposite, 2/1. Thus in this moment an abstract, mathematical event is precisely linked with a physical, sensory perception; our direct, intuitional response to this phenomenon of sound (the octave) coincides with its concrete, measured definition.

Hence we experience in this auditory perception a simultaneous interwovenness of interior with exterior, and we can generalize this response to invoke the possibility of a merger of intuitional and material realms, the realms of art and science, of time and space. There may be another such moment in the created world, but the Pythagoreans did not know of it, nor do we. This is the essential spirit of the perception of Harmony, and for the Pythagoreans it was the only true supernatural moment: a tangible experience of the simultaneity of opposites. It was considered to be true Magic, an omnipresent and authentic mystery.

13

It was by means of geometry that the Pythagoreans poised themselves at this unique transition where heard vibration becomes seen form; and their geometry, as we shall see, explores the relationships of musical harmony. Although interwoven in function, our two major intellectual senses, sight and hearing, use our intelligence in two completely different ways. For example, with our optic intelligence, in order to form a thought we make an image in our mind. Hearing, on the other hand, uses the mind in an immediate, unimaged response whose action is expansive, evoking a response from the emotive centres. Nowadays this emotive, sound-sensing faculty is usually associated with subjective, emotional, aesthetic or spiritual experiences. We tend to forget that it is also involved when the reason perceives invariant relationships. Therefore when we place the auditory capacity at the centre of our sensory experience we can become aware that it is possible to listen to a colour, or to a movement. This intellectual capacity is quite different from the 'visual', analytical and sequential one we normally employ. It is this capacity, which is associated with the right hemisphere of the brain, that recognizes patterns in space, or wholes of any kind. It can perceive opposites in simultaneity and grasp functions which to the analytic faculty appear irrational. It is in fact the perfect complement of the 'left hemisphere', visual, analytic capacity, for it absorbs spatial and simultaneous orders while the 'left' rational faculty is best suited to grasp temporal, sequential organization. The esoteric, *functional* aspect of Number, for instance, would be apprehended through the 'right hemisphere' faculty, while the exoteric, enumerative aspect of Number is apprehended by the 'left'.

This innate intellectual quality resembles very closely what the Greeks called Pure Reason, or what in India was called the 'heart-mind'. The ancient Egyptians had a beautiful name for it, the Intelligence of the Heart, and to achieve this quality of understanding was life's implicit goal. The practice of Geometry, while also utilizing the analytic faculty, uses and cultivates this audial, intuitive aspect of mind. For example, one experiences the fact of geometric growth through the image of the square with its diagonal which forms the side of a second square. This is an unreasoned certainty absorbed by the mind from the actual experience of executing the drawing. The logic is contained within the lines on paper, which cannot be drawn in any other way.

As geometers, equipped only with compasses and straight-edge, we enter the two-dimensional world of the representation of form. A link is forged between the most concrete (form and measure) and the most abstract realms of thought. By seeking the invariable relationships by which forms are governed and interconnected we bring ourselves into resonance with universal order. By re-enacting the genesis of these forms we seek to know the principles of evolution. And by thus raising our own patterns of thought to these archetypal levels, we invite the force of these levels to penetrate our mind and thinking. Our intuition is enlivened, and perhaps, as Plato says, the soul's eye might be purified and kindled afresh 'for it is by it alone that we contemplate the truth'.

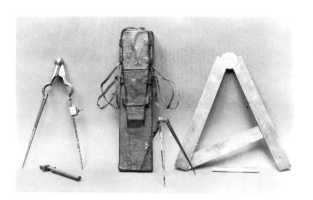

'Numbers are the sources of form and energy in the world. They are dynamic and active even among themselves . . . almost human in their capacity for mutual influence.' (Theon of Smyrna.) Numbers, in the Pythagorean view, can be androgynous or sexual, procreators or progeny, active or passive, heterogeneous or promiscuous, generous or miserly, undefined or individualized. They have their attractions, repulsions, families, friends; they make marriage contracts. They are in fact the very elements of nature. The tools of geometry and number represent the means to attain knowledge of both external and internal space and time. These instruments, once used by architects and philosophers, became instead, from the time of the 'Age of Reason', the tools of the engineer.

It seems to be the basic assumption of traditional philosophies that human intellectual powers are for the purpose of accelerating our own evolution beyond the restraints of the biological determinism which binds all other living organisms. Methods such as yoga, meditation, concentration, the arts, the crafts, are psycho-physical techniques to further this fundamental goal. The practice of Sacred Geometry is one of these essential techniques of self-development.

Each of the diagrams in the small squares represents a different system or technique of thought for understanding the world and its structures. The first task of the spiritual aspirant confronting the varied contemplative paths is to harmonize the five universal constituents which compose his body (earth, air, fire, water and *prana*). His clear cognition of the outer and inner worlds is dependent upon the harmonious accord which he establishes between these elemental states in his own body and these same elements in nature. Each geometric cosmogram is meant to assist him in these attempts at liberation through harmonization.

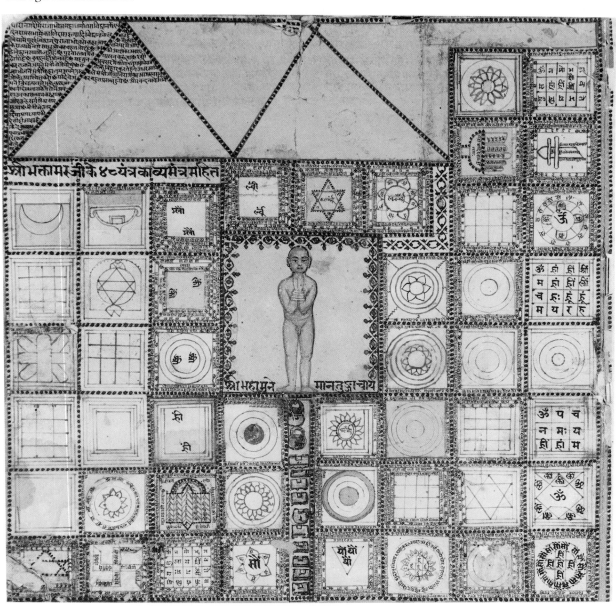

II Sacred Geometry: Metaphor of Universal Order

Whether the product of an eastern or a western culture, the circular mandala or sacred diagram is a familiar and pervasive image throughout the history of art. India, Tibet, Islam and medieval Europe have all produced them in abundance, and most tribal cultures employ them as well, either in the form of paintings or buildings or dances. Such diagrams are often based on the division of the circle into four quarters, and all the parts and elements involved are interrelated into a unified design. They are most often in some way cosmological; that is, they represent in symbol what is thought to be the essential structure of the universe: for example, the four spatial directions, the four elements, the four seasons, sometimes the twelve signs of the zodiac, various divinities and often man himself. But what is most consistently striking about this form of diagram is that it expresses the notion of *cosmos*, that is of reality conceived as an organized, unified whole.

Ancient geometry rests on no *a priori* axioms or assumptions. Unlike Euclidian and the more recent geometries, the starting point of ancient geometric thought is not a network of intellectual definitions or abstractions, but instead a meditation upon a metaphysical Unity, followed by an attempt to symbolize visually and to contemplate the pure, formal order which springs forth from this incomprehensible Oneness. It is the approach to the starting point of the geometric activity which radically separates what we may call the sacred from the mundane or secular geometries. Ancient geometry begins with *One*, while modern mathematics and geometry begin with *Zero*.

One of the most striking uses of the mandala is dome architecture, Islamic and Christian. The square represents the earth held in fourfold embrace by the circular vault of the sky and hence subject to the ever-flowing wheel of time. When the incessant movement of the universe, depicted by the circle, yields to comprehensible order, one finds the square. The square then presupposes the circle and results from it. The relationship of form and movement, space and time, is evoked by the mandala.

Here the mandala of Unity is inscribed on the hand, held in a ritual gesture, of a Japanese Buddhist deity. The mandala is the division of the Unity circle into the comprehensible forms of square, hexagon, octagon, enneagon, etc., and these forms are considered to be the primary thoughts of God emerging out of the circular Unity. But for thoughts to become activities and workings they require a will or force of intention, which is symbolized by the hand. The positions of the hand can be systematized to form a medium of communication (*mudra*) in which the gesture mirrors the various forces through which the dispositions of creative mind come into manifest form.

   I would like to consider in some detail these two symbolic beginnings, One and Zero, because they provide an exceptional example of how mathematical concepts are the prototypes for the dynamics of thought, of structuring and of action.

   Let us first consider zero, which is a relatively recent idea in the history of thought, yet is already so ingrained in us that we can hardly think without it. The origins of this symbol date back to sometime before the eighth century AD, when we have a record of its first written appearance in a mathematical text from India. It is interesting to note that during the century just prior to this time a particular line of thought had begun to develop in India which found its expression in both Hinduism (through Shankhara) and Buddhism (through Narayana). This school laid exclusive emphasis on the goal of obtaining personal transcendence and escape from *karma* through renunciation of the natural world, even to the extent of mortification of the physical body. The goal of this highly ascetic pursuit was the attainment of an utterly impersonal, blank void, a total cessation of movement within consciousness. A description of it attributed to Buddha is 'a state of incognizable, imperishable, selfless absence'. This single aspect or possibility of meditative experience was held to be the ultimate goal of the created Universe as well as the goal of all individual spiritual development. In retrospect this is now considered by many to be a dark period in the long, rich spiritual heritage of India, a decline from the previous tradition which upheld a spiritual significance in both the manifested and the unmanifested expressions of God, and whose tantric and yogic practices worked towards an intensification of the relationship and harmonization between matter and spirit. It was at this time that the concept of zero took on a new tangibility and

The primary geometric forms are considered to be the crystallizations of the creative thoughts of God, and the human hand, in manipulating and constructing these forms, will learn to position itself in the essential poses of gesture-language.

17

presence. The result was that it achieved a specific name and symbol in both metaphysics and mathematics. In mathematics it came to be considered just like the other numbers, as a symbol which can be operated upon and calculated with. The name given to this concept in Sanskrit was *sunya*, meaning 'empty'.

Some mathematical historians will argue that the exclusive claim of the Hindu notion of zero is not verifiable, claiming that before India, in Babylonia, Greece and in the Maya civilization a symbol was sometimes used to denote an empty column. In a number such as 203, for instance, the empty column is where the zero is. In Babylonia the empty space would be designated by two marks like this //, and in Greece by a small O with a dash, and the Maya used a sort of egg-shaped symbol. But to mark an empty column is only a notational procedure, while on the other hand in Indian mathematics the zero is treated as a tangible entity, as a number. The Indian mathematicians wrote such things as $(a \times 0) \div 0 = a$. Aristotle and other Greek teachers had talked about the concept of zero philosophically, but Greek mathematics, fortified as it was by the Pythagorean teaching from Egypt, resisted the incorporation of zero into its system.

The Arabs, who functioned from the ninth to the fourteenth centuries as the transmitters of knowledge and culture from the ancient, declining cultures of the far east and Egypt, carried this knowledge into the emerging ferment of Western Europe. During these centuries they picked up the concept of zero along with nine other numerical symbols which had developed in India. The less mystical and more practical orientation of the Arab mentality saw in these symbols a practical device for facilitating calculation and recording large numbers, particularly numbers containing an empty column, such as 1505.

Roman numerals, in use right through the Middle Ages, maintained a notation similar to that of Egyptian numeration in that both were based on groupings which did not require a zero to indicate the empty column:

Egyptian   𐦀 𝑒𝑒 ''' = 1505
                 𝑒𝑒𝑒 ''
Roman   MDV   = 1505

Each unit of growth, the tens, the hundreds, the thousands, etc., had a separate symbol, giving a decimal system with no zero.

The great eighth-century Arab mathematician, Al-Khwrizmi, carried the Indian numerals with zero to the Islamic world. Then another 400 years passed before the works of Al-Gorisma (whose name became the basis of our word algorithm) were brought into Europe through the Arabic settlements in Spain. His works were translated into Latin somewhere in the twelfth century. Gradually this 'Arabic' number system was introduced into medieval Europe and began to support radical changes in Western science and thought.

Some of the monastic Orders resisted the adoption of this system of decimal notation with zero, claiming particularly that zero was a device of the Devil. Among those who refused it was the Cistercian Order whose mystic and gnostic philosophy was the inspiration and foundation for the construction of the Gothic cathedrals, the cosmic temples of the Piscean Age. But the merchants adopted the Arabic numerals and zero because they gave a mechanical ease to calculating operations and the recording of quantities. It was then through the mercantile impulse that zero took root.

The consequences were enormous. First of all, within the structure of arithmetic itself, the additive basis of calculation had to be cast aside. Formerly the addition of one number to another number always produced a sum larger than either of the original numbers. This was of course nullified by the utilization of zero. Other laws of arithmetic were also altered, so that we are now able to have operations such as the following:

$$3+0 = 3$$
$$3-0 = 3$$
$$03 \quad = 3$$
$$30 \quad = 3 \times 10$$
$$\text{but} \quad 3 \times 0 = 0$$
$$\text{and} \quad 3 \div 0 = 0 \ (???)$$

Here logic completely breaks down. The illogic of the symbol was accepted because of the convenience it afforded to quantitative operations. Yet this breakdown of the simple, natural logic of the arithmetic structure allowed a complicated mental logic to take its place and invited into mathematics a whole range of numerical and symbolic entities, some of which have no verifiable concept or geometric form behind them. Arising from the sixteenth century onwards, these entities include relative numbers (i.e. negative quantities such as -3); infinite decimal numbers; algebraic irrational numbers such as the cube root of 10; transcendental irrational numbers (numbers such as e, the basis of logarithms, which satisfy no rational algebraic equation); imaginary numbers such as the square root of -1; complex numbers (the sum of a real number and an imaginary number); and literal numbers (letters representing mathematical formulae). The invention of zero permitted numbers to represent ideas which have no form. This signals a change in the definition of the word 'idea', which in antiquity was synonymous with 'form', and implies geometry.

The theological impulse of the Indian mentality did not allow it to place zero at the beginning of the series. Zero was placed after 9. It was not until the late sixteenth century in Europe, the dawn of the Age of Reason, that 0 was placed before 1, allowing for the concept of negative numbers.

Not only has zero become indispensable in the mathematical system on which our science and technology depends, but it has become implicitly translated into our philosophy and theologies, our way of viewing nature, our attitudes towards our own natures and toward the environment. We have seen how in India the adoption of zero was associated with a doctrine which negated the reality of the material world. The Sanskrit name for zero, *sunya*, meaning 'empty', became *chiffra* in Latin, which carries the meaning of null or nothing. Needless to say, 'nothing' is a different concept from 'empty'. Also at this period in India the Sanskrit word *maya* took on a new meaning. Originally it meant 'the power to divide' or the 'dividing mind', but at this time it came to mean 'illusion', or the material aspect of the universe as illusion. We can see the reverse of this spiritual nihilism in the materialism of the West after the Industrial Revolution, when the spiritual aspect of reality came to be seen as illusory.

The 'western' rationalistic mentality negated the ancient and revered spiritual concept of Unity, for with the adoption of zero, Unity looses its first position and becomes merely a quantity among other quantities. The advent of zero allows one to consider anything below the quantitative number series as nil or of no account, while anything beyond the quantitatively comprehensible range becomes an extrapolation subsumed under the word God and deemed religious or superstitious. Hence zero provided a framework in western thinking for the development of atheism or negation of the spiritual.

From the point of view of the natural world, zero does not exist; it is a completely mental entity. Yet the impact of this symbol was so great that it caused the supposedly empirical physics of the nineteenth century to adapt an atomic theory in which matter was modelled as composed of tiny building blocks, little spheres floating in a zero-empty void. Zero continued to be formative in the nineteenth-century world view through the idea that there is a separation between the quantitative and the non-quantitative; the extreme degree of this idea was that everything which is non-quantitative is non-existent or zero. Twentieth-century nuclear physics no longer conceives of the atom as a separate attracting and repelling particle, but instead it poses a field or matrix of interconnected, continually transforming energy fields of particles and patterns. Particles indistinguishable from process; matter indistinguishable from events. Likewise in the heavens, what was once thought to be a black, empty void with bodies floating in it is now known to be filled with substance-energy. Between a stellar body and the region surrounding it, there is a field continuum of which the star-body is simply a densification. While weaning us from the nineteenth-century world view, both microcosmic and macrocosmic, today's science shows us a continual fluctuation and alternation between matter and energy, confirming that in the natural world there is no zero.

The notion of zero also had its effect on our psychological conceptualizations. Ideas such as the finality of death and the fear of it, the separation of heaven and earth, the whole range of existential philosophies based on the despair and absurdity of a world followed by non-being, all owe much to the notion of zero. We saw ourselves as separate individuals moving in a space which was other than ourselves, encountering in that space other beings separate from and other than ourselves. But these concepts are now also loosing their hold. We know now that we exist in groups, determined by various levels of energetic affinities, repelling, exchanging and absorbing through interconnected, subtle energetic communications. And our being extends outward through various energy fields to connect with larger fields. We have had to learn that there is nowhere that we can dispose of the things we have finished using – that there is no zero drain in our sink; there is no factory pipe or hole in the ground that does not lead somewhere. Everything remains here with us; the cycles of growth, utilization and decay are unbroken. There is no throw-away bottle.

With zero we have at the beginning of modern mathematics a number concept which is philosophically misleading and one which creates a separation between our system of numerical symbols and the structure of the natural world. On the other hand, with the notion of Unity which governs ancient mathematics, there is no such dichotomy.

The notion of Unity remains, literally, unthinkable; simply because in order for anything to be, to exist, it must, in the very positive affirmation of itself, negate that which it is not. Cold is only cold because it is the negation of heat. For a thing to be, its opposite must also be. There is then at the beginning of the created world a contingency of *division of Unity* into two. With two, number begins. This same law governs our understanding, for in order to comprehend any objective state we must acknowledge and negate its opposite. R.A. Schwaller de Lubicz says,

> The Number One is only definable through the number two: it is multiplicity which reveals unity. . . . The intelligence of things exists only through what we may call an original fractioning and the comparison of these fractions to one another, which is then only an enumeration of the aspects of Unity.

Thus, unthinkable though Unity may be, both reason and spiritual experience compel the traditional thinker to place it at the beginning. Everything that exists in his mathematical problem or in his universe is a fraction of the unknown One, and because these parts can be related proportionately to one another they are knowable. Sri Aurobindo says,

> At the origin of things we are faced with an infinite containing a mass of un-explained finites; an indivisible full of endless divisions, an immutable teeming with mutations and differentiations, a cosmic paradox is at the beginning of all things. This paradox can only be explained as One, but this is an infinite Oneness which can contain the hundred and the thousand and the million and the billion

and the trillion. . . . This does not mean that the One is plural, or can be limited or described as the sum of the many. On the contrary, it can contain the infinite many because it exceeds all limitation or description by multiplicity, and exceeds at the same time all limitation by finite, conceptual oneness.

(The Life Divine.)

Unity is a philosophic concept and a mystic experience expressible mathematically. The Western mentality, however, withdrew its discipline of acknowledging a supra-rational, unknowable mystery as its first principle. But in rejecting this reverence to a single unknowable unity, our mathematics and science developed into a system requiring complex, interconnected hypotheses, imaginary entities such as those mentioned above, and unknown x quantities which must be manipulated, quantified or equalized as in the algebraic form of thought. So the unknown appears not just once but at every turn, and can be dealt with only by seeking quantitative solutions.

Our present thought is based on the following numerical and logical sequence:

$$-5, -4, -3, -2, -1, 1, 0, 1, 2, 3, 4, 5$$

With zero in the centre, there is a quantitative expansion 1, 2, 3 . . . and our sense of balance requires having $-1, -2, -3$ . . . on the other side, giving a series of non-existent abstractions (negative quantities) which demand an absurd logic. The system has a break-point, zero, disconnecting the continuum and dissociating the positive numbers from the negative balancing series.

In the ancient Egyptian numerical progression, beginning with one rather than zero, all the elements are natural and real:

$$1/5, 1/4, 1/3, 1/2, 1, 2, 3, 4, 5$$

All the elements flow out from the central unity in accordance with the law of inversion or reciprocity. The Egyptians based their mathematics on this simple, natural series of numbers, performing sophisticated operations with it for which we now need complex algebra and trigonometry. We have already seen the natural demonstration of this series in the physical laws of sound. The plucked string, when divided in half, produced double the frequency of vibrations. Thus this series expresses the essential law of Harmony.

Much of post-Einsteinian physics seems to have this poise of mind as its basis, as inversion plays a major role in Relativity Theory, the Uncertainty Principle, and in such concepts as that of Black Holes. The idea of a continual interchange between matter and energy also requires this poise.

Such metaphysical concepts as the immortality of the soul, rebirth and reincarnation are also more fully grasped by means of the notion of reciprocity. To the Egyptians, the nether world to which the soul proceeded after death was called the 'inverted world', the *Dwat*. The progression of inverse (reciprocal) elements supplies a mental basis for the notion of perpetual interchange through reversal.

The idea of the unknowable Unity at the beginning has been the basis of many philosophies and mythological systems. While Shankhara, with the Buddhism of a certain period, posited the void as a fundamental assumption, the main stream of Hinduism has always rested on the notion of the One, the Divine, who divided himself within himself to form his own self-created opposite, the manifested universe. Within the divine self-regard, three qualities of himself became distinguished: *Sat* (immobile being), *Chit* (consciousness-force) and *Ananda* (bliss). The original unity, represented by a circle, is then restated in the concept of the Real-Idea, the thought of God, which the Hindus called the *bindu* or seed, what we call the geometrical point. The point, according to the *Shiva Sutra Vimarshini Com-*

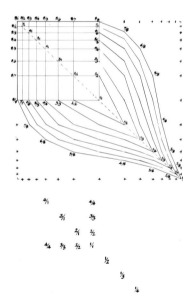

The natural progression of whole numbers with their inverse progression is a pattern for the formation of the most common leaf form.

Music is pervaded by the fundamental law of reciprocity; changes in frequency and wave length are reciprocal. Rising or falling tones, as reciprocal arithmetic ratios, are applied to string-lengths. 'Major' and 'minor' are reciprocal tonal patterns. As Ernest McClain points out in *The Myth of Invariance*, Plato conceived the World-Soul as constituted of reciprocal ratios identical with those which, in Hindu mythology, create the musical 'drum of Shiva', the pulsating instrument of creation (see p. 81).

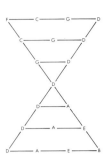

mentaries, forms the limit between the manifest and non-manifest, between the spatial and the non-spatial. The *bindu* corresponds to the 'seed-sound idea' of the Tantras. The Divine transforms himself into sound vibration (*nada*), and proliferates the universe, which is not different from himself, by giving form or verbal expression to this self-idea. Ramakrishna summarized the scripture by saying, 'The Universe is nothing but the Divine uttering his own name to himself.'

Thus the universe springs forth from the Word. This transcendent Word is only a vibration (a materialization) of the Divine thought which gives rise to the fractioning of unity which is creation. The Word (*saabda* in Sanskrit, the *logos* of the Christians and Gnostics), whose nature is pure vibration, represents the essential nature of all that exists. Concentric vibrational waves span outward from innumerable centres and their overlappings (interference patterns) form nodules of trapped energy which become the whirling, fiery bodies of the heavens. The Real-Idea, the Purusha, the inaudible and invisible point of the sound-idea remains fixed and immutable. Its names, however, can be investigated through geometry and number. This emitted sound, the naming of God's idea, is what the Pythagoreans would call the Music of the Spheres.

In ancient Egypt the primordial vibrational field (called *nada* in India) is called Nun, the primal ocean. It is the One imaged as undifferentiated cosmic substance, the source of all creation. Submerged within this primal ocean is Atum, the creator, who must first distinguish himself from Nun in order for creation to begin. Atum is masculine, and analogous to *Chit* (consciousness-force) of the Indian myth. Atum is pictured in a state of total self-absorbed bliss. Some versions of the myth say that Atum is masturbating. His blissful self-contemplation provokes his ejaculation and this ejaculation catches in his throat, causing him to cough his own seed out of his mouth. He coughed and spit out Shu and Tefnut, who, together with himself, form the first triad of the nine great *Neteru* or principles of creation.

Let us note the relationship of this creation myth to the Egyptian mathematical notation in which fractions are represented by drawing a mouth as the numerator and unit marks underneath for the denominator, imaging the idea of seed-powers being emitted from the mouth, the creative Word ⟨⟩ = 1/3. The hieroglyphic sign for the mouth ⟨⟩ is the same sign used to write the name of the supreme being, *Rê* (who, as creator, is known as Atum-Rê). Atum's projected seed enters into the primal vibration of Nun and coagulates it into the universe of forms, just as the sperm coagulates the albuminous substance of the ovum. (This and other functional correlations with Egyptian myth have been developed by Lucie Lamy in *Egyptian Mysteries*.)

Notice that both this 'mouth' symbol from Egypt and the path of a vibrating string have a flattened, vesical form.

Today, in the field theory of modern astro-physics, the universe is conceived as an integral, incomprehensibly vast vibrating field of ionized, pre-gaseous plasma, an image not unlike that of the *Nun* or cosmic ocean of the Egyptian myth, or the *Prakriti* of the Hindu cosmology. Within this field gravitational influences are triggered which cause a warp and densification into nodal patterns. The dis-equilibrium and turbulence caused by the newly formed galactic mass-centres under the forces of contraction releases compound ripples causing violent, abrupt changes in the pressure and density of the whole cosmic plasma. These are referred to as galactic 'sonic booms', sonic because indeed the propagation of any sound is simply the rapid oscillatory pressure-density change in any medium. These whirling sonic shocks create a spin in the entire galactic cloud and within the inner regions set up by this spin the stars are born. This clearly restates the ancient image of universal creation through sound waves or the Word of God; science reaffirms that visible stars and galaxies are spiral blast patterns, residual imprints of standing shock waves from the thundering voice of the Universe.

Thus the most recent scientific model of creation is allied to the image given in ancient mythology, and both acknowledge an absolute singularity or Unity at the beginning. In terms of the orthodoxy of ancient mathematics, the symbols of mathematics should reflect the realities they describe. With zero and the army of merely mental and statistical signs which followed from it, we are very far from having a system of mathematical symbols which corresponds to the pure geometric order of living space.

III The Primal Act: The Division of Unity

Those who use geometric figures to describe the beginning of Creation must attempt to show how an absolute Unity can become multiplicity and diversity. Geometry attempts to recapture the orderly movement from an infinite formlessness to an endless interconnected array of forms, and in recreating this mysterious passage from One to Two, it renders it symbolically visible.

From both the metaphysical and natural points of view it is false to say that in order to arrive at two, you take two ones and put them together. One only need look at the way in which a living cell becomes two. For One by definition is singular, it is Unity, therefore all inclusive. There cannot be two Ones. Unity, as the perfect symbol for God, divides itself from within itself, thus creating Two: the 'self' and the 'me' of God, so to speak; the creator unity and the created multiplicity.

Unity creates by dividing itself, and this can be symbolized geometrically in several different ways, depending upon how the original Unity is graphically represented. Unity can be appropriately represented as a circle, but the very in-commensurability of the circle indicates that this figure belongs to a level of symbols beyond reasoning and measure. Unity can be restated as the Square, which, with its perfect symmetry, also represents wholeness, and yields to comprehensible measure. In geometrical philosophy the circle is the symbol of unmanifest Unity, while the square represents Unity poised, as it were, for manifestation. The square represents the four primary orientations, north, south, east and west, which make space comprehensible, and it is formed by two pairs of perfectly equal yet opposi-tional linear elements, thus graphically fulfilling the description of universal Nature found in Taoist and other ancient philosophies.

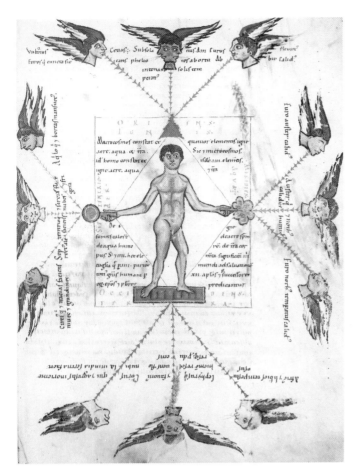

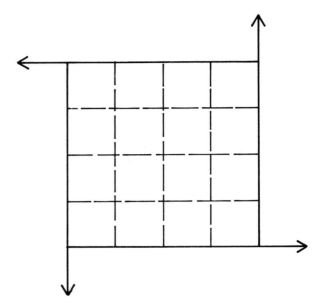

The square is the result of a crossing.

The four orientations were related to the four constituents of creation: earth, air, fire, water.

By definition the square is four equal straight lines joined at right angles. But a more important definition is that the square *is* the fact that any number, when multiplied by itself, is a square. Multiplication is symbolized by a cross, and this graphic symbol itself is an accurate definition of multiplication. When we cross a vertical with a horizontal giving these line-movements equal units of length, say 4 for example, we see that this crossing generates a square surface: a tangible, measurable entity comes into existence as a result of crossing. The principle can be transferred symbolically to the crossing of any contraries such as the crossing of male and female which gives birth to an individual being, or the crossing of warp and weft which gives birth to a cloth surface, or the crossing of darkness and light which gives birth to tangible, visible form, or the crossing of matter and spirit which gives birth to life itself. So the crossing is an *action-principle* which the square perfectly represents.

The word Nature means 'that which is born', and all birth into nature requires this crossing of opposites. So the square came to represent the earth, and as such symbolized the conscious experience of finite existence, of what is born into Nature. This brings us to the problem of whether the sides of the square are curved or straight: if the overall reality of the universe is an endless curvature, an endless movement, there is yet a consciousness which is capable of temporarily arresting, both conceptually and perceptually, segments of the universal continuum. This objective consciousness might be seen as a reduced velocity of the universal consciousness, and has as its instrument the cerebral cortex in man. The Indians called this power of isolation and arrestation of the ever-moving universal Becoming *tapas*. The Greek philosopher Heraclitus likened it to a paralysis of vision such as one experiences when stung by a scorpion. He called objectivization the 'scorpion sting'.

The Buddhist and Hindu philosophers were concerned lest human consciousness become fascinated or preoccupied by this segmented perception of reality. To use a familiar Buddhist analogy, Time is like a necklace of square beads of tangible objects, or moments or events, and to be absorbed by this succession of limited frames is *maya* or illusion, whereas only the inner thread of the necklace, the unimaginable continuum, is reality.

Pythagoras, however, taught that the experience of life in a finite, limited body was specifically for the purpose of discovering and manifesting supernatural existence within the finite. One's concentration, then, should also be on the finite itself, to discover how this finite could contain intrinsically a power to express the infinite. This does not mean concentration on finite, material effects, but on the abstract principles revealed in the finite world, and the Causes which create and support this embodiment. Hence Pythagorean mathematics were limited to whole numbers, that is, definable, arrested states, and sought after universal expressions within the measurable, geometric frame of the square, a profound symbol of finite perfection.

The following Workbook is the first of nine such sections in this volume, intended to take readers step by step through the principal drawings and concepts of Sacred Geometry. It is suggested that readers take compasses and straight edge and draw for themselves, following the instructions given adjacent to the drawings, each of the figures in the Workbook sections. It is also advisable to use graph paper for these drawings, so that verification for certain relationships can be obtained by simply counting the grid squares.

Workbook 1

The square cut by its diagonal; $\sqrt{2}$

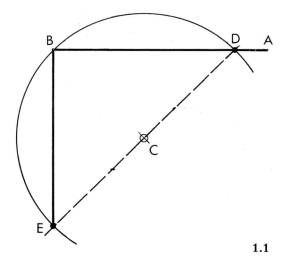

1.1

Drawing 1.1. Draw any line *AB* and locate any point *C* below *AB* and somewhere near its midpoint. With centre *C* and radius *CB* swing an arc of at least half a circle, cutting *AB* at *D*. Join *C* and *D* continuing the line until it cuts the arc at *E*. Draw *EB* perpendicular to *AB*.

Drawing 1.2. With centre *B* and radius *BA* swing an arc until it crosses *BE* at *G*. From centres *G* and *A* and radius *AB* swing two arcs intersecting at *F*. Draw square *ABGF*.

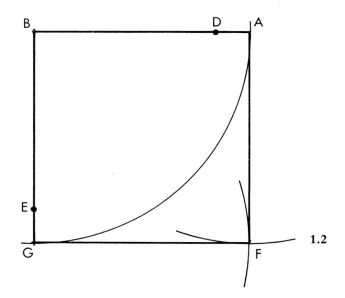

1.2

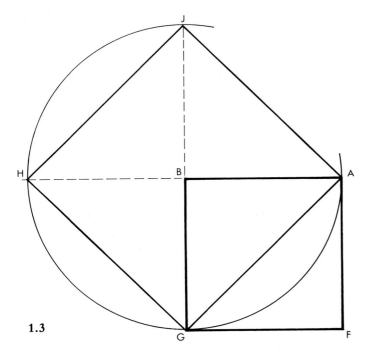

1.3

Drawing 1.3. Within square *ABGF* draw diagonal *AG*. Using the same method as in Drawing 1.1 construct a line perpendicular to *AG* at *G*. With *B* as centre and *BA* as radius, swing an arc of at least half of a circle to determine points *H* and *J*. Using the same method as in Drawing 1.2, complete the square *AGHJ*.

The side of square *AGHJ* (square 2) is exactly equal to the diagonal of square *ABGF* (the primary square).

The area of square 2 is exactly twice that of the primary square. (This is intuitively evident by the larger square containing four identical triangles, whereas the primary square contains only two.)

The side of a square is called its root ($\sqrt{}$). The side of the primary square (square 1) is $\sqrt{1}$, and that of square 2 is $\sqrt{2}$.

The diagonal of square 2 is equal to 2, exactly twice the side of the primary square.

This relationship can be written as:

$$\frac{\text{root}}{\text{diag}} : \frac{\text{root}}{\text{diag}} :: \frac{1}{\sqrt{2}} : \frac{\sqrt{2}}{2}$$

but it can also be considered:

$$\frac{\text{root}}{\text{root}} : \frac{\text{diag}}{\text{diag}} :: \frac{1}{\sqrt{2}} : \frac{\sqrt{2}}{2}$$

$$\frac{\text{root}}{\text{diag}} : \frac{\text{diag}}{\text{root}} :: \frac{1}{\sqrt{2}} : \frac{\sqrt{2}}{2}$$

These relationships seem to be a logical paradox, but if the reader studies the drawing he will find they are geometrically true. Even as the squares increase in size, their root-diagonal relationships remain proportional identities.

Drawing 1.4. Repeat the process of Drawing 1.3. With centre *J* swing an arc equal to the side of square 2. Extend the sides *AJ* and *HJ* until they intersect the arc at *K* and *M*. Draw square 3, *MKHA*. In a similar manner construct squares 4, 5, etc.

The relationship of the side to the diagonal of each square, and of each square to the next larger square, is identical to that of square 1 to square 2. This can be written as:

$$\frac{1}{\sqrt{2}} : \frac{\sqrt{2}}{2} : \frac{2}{2\sqrt{2}} : \frac{2\sqrt{2}}{4} : \frac{4}{4\sqrt{2}} : \frac{4\sqrt{2}}{8} \text{ etc.}$$

or, in general terms,

$$\frac{a}{b} : \frac{b}{c} : \frac{c}{d} : \frac{d}{e} : \frac{e}{f} \text{ etc.}$$

This type of progression is called a 'geometric progression', where the numerator, when multiplied by the denominator of the second relationship is equal to the multiplication of the numerator of the second relationship by the denominator of the first relationship. This law of cross multiplication between sets of numerators and denominators holds true for any ratios in the progression, whether in sequence or not.

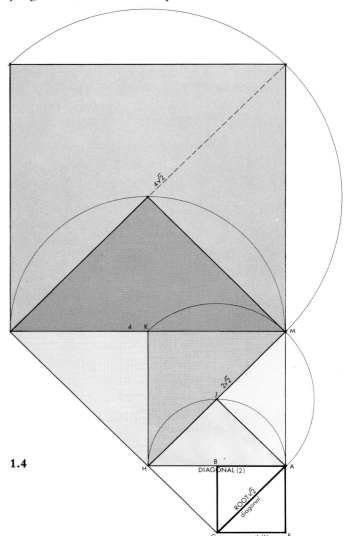

1.4

Drawing 1.5. This represents a variation of the previous geometric progression, but extended in the opposite direction of diminution. Given square *ABCD*, draw the diagonals *DB* and *AC*. With *B* and *C* as centres and radius *EB* equal to half the diagonal, swing two arcs intersecting at *F*. Draw line *EF*, intersecting the sides of square 1 at *G*. With *B* and *F* as centres and radius *GF*, swing two arcs intersecting at *H*. Draw square *BHFG* (square 2). Repeat this process, constructing squares which progressively diminish according to the geometric progression, 2, 4, 8, 16, 32 etc.

In both examples the square and its diagonal express the creation of Two from Unity (the initial square), and a consequent proliferation of number through a geometric sequence.

The square divided by its diagonal provides an archetypal model for geometric proportions and progressions of this type, that is $1:\sqrt{2}::\sqrt{2}:2$, where each term (or ratio) is multiplied by a constant value in order to achieve the next term in the progression. A fixed, proportional increase or rate can be the generative pattern for other infinitely expanding geometric progressions, for example, $1:\sqrt{3}::\sqrt{3}:3$, or $1:3::3:9::9:27 \ldots$ etc. (see p. 35). In this geometric demonstration of the

relationship between proportion and progression we are reminded of the alchemical axiom that everything in creation is formed from a fixed, immutable component (proportion) as well as a volatile, mutable component (progression).

The relationship between the fixed and the volatile (between proportion and progression) is a key to Sacred Geometry: everything which is manifest, be it in the physical world or in the world of mental images and conceptions, belongs to the ever-flowing progressions of constant change; it is only the non-manifest realm of Principles which is immutable. Our science errs in attempting to attach fixed, absolute laws and definitions to the changing world of appearances. The history of science shows us perpetually discarding or revising one world model after another. Because of the disturbingly unstable quality of scientific knowledge, not only our physicists, but also our philosophers, artists and society as a whole have become relativists. But the unchanging, generative principles remain, and our contemporary rejection of them is taking place only because we have sought for the permanent in the empirical world instead of in its true abode, the metaphysical.

1.5

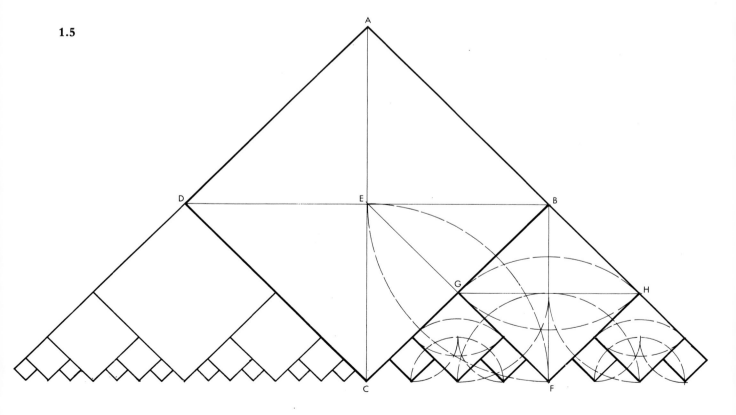

In Drawing 1.3 we witnessed the division of Unity through the drawing of the square's diagonal. The side of the original square, called its 'root', is given the value of 1 since it is the first or primary unit. The area of this square is also 1 because $1 \times 1 = 1$. The simple act of drawing the diagonal has given rise to 2, *not because the square has been divided in half*, but because square 2 is implied, since the diagonal of square 1 is the root of square 2, and square 2 is exactly *double in area* to square 1.

The reader may justifiably wonder why, having arrived at the symbol of the square, we must further consider the square built on its diagonal; for that matter, why consider this diagonal at all? This requires that we define the cause-effect relationship as seen in contemplative geometry. Once the four-cornered square has been drawn, one has implicitly all that is necessary to draw the square's diagonal lines. Moreover this diagonal line (like any straight line) is implicitly the side or root of a square. In other words, we are obliged to think through or make explicit that which is implicit in any geometric figure. A form is a geometric system and like any system, biological, chemical or other, it must be seen in the unfolding continuum of its components in their cause-effect relationships. The movement from implicit to explicit is similar to that from cause to effect. It is only in the arbitrary mental world that cause can be separated from effect, while in the natural world they are inseparable: a cause is not a cause unless it has an effect. If we carry this logic further we see that the square surface also only exists in a continuous relationship to a cubic volume, of which it forms one of the six faces. In contemplative geometry the attempt is always to follow the complete movement from the purely abstract, two-dimensional world of line, then plane, as it becomes explicit in the actual world of three-dimensional volume.

To return to our square, two paradoxes have been revealed in the act of its division by the diagonal. The first lies in the uncanny coinciding of the two functions of root and diagonal in the geometric moment of the square root of 2. The same line unit is both root and diagonal, the paradox of sameness and difference. This simultaneity of function yields three seemingly contradictory yet geometrically true relationships:

$$\frac{root}{diagonal} : \frac{diagonal}{root} \qquad \frac{root}{root} : \frac{diagonal}{diagonal} \qquad \frac{root}{diagonal} : \frac{root}{diagonal} = \frac{1}{\sqrt{2}} : \frac{\sqrt{2}}{2}$$

The second paradox lies in the fact that the *half* (the square halved by the diagonal) produces the *double*, as in the generation of musical tone and in the mystery of biological growth from cellular division.

The square root of 2 is an irrational function and a universally applicable relationship. Since everything in the natural world undergoes change, this root, being invariant, is by definition supernatural or supra-rational, that is to say it is a symbol of the archetypal realm. The Pythagoreans are said to have referred to the incommensurable numbers as 'unutterable'. We can be assured that it was neither secrecy nor puerile piety which led them to so designate them. It was, on the contrary, the keen discretion of an intellect aware of and maintaining a relationship between Number and cosmic realities.

Drawing 1.4 shows how the creation of 2 leads to endless proliferation through the geometric progression $a:b::b:c$, etc., or, expressed numerically, $1:2::2:4::4:8::16:32::32:64$, etc. No matter how vast the numerical relationships become, this proportion $a:b::b:c$ remains unchanged. This progression can extend toward diminution as well as toward vastness through a bisection of the square, concurrent with a numerical expansion through the power of the diagonal of the square. The square root of 2 thus represents the *power of multiplicity* which can extend itself both towards unlimited expansion and towards utterly minute finiteness. This figure perfectly represents the growth pattern of cellular fission in living organisms. Not only number but form proliferates from the division of Unity.

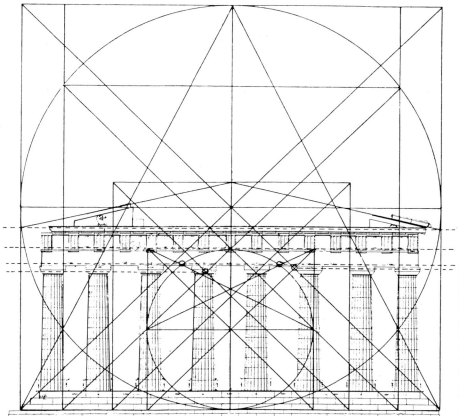

In this geometric analysis of the Parthenon by Tons Brunés from his book *The Secrets of Ancient Geometry*, it can be seen that the architecture is governed by the relationship between side and diagonal in a series of squares. Each of the squares is in relationship to the larger square enclosing it in the ratio of 1 to 1·25; therefore the whole proportional system is based on the functional relationship of √2 to 1 to 1·25 (or 5/4).

When we speak of roots of squares and roots of cubes we are using a very ancient designation which associates this mathematical function with the vegetal root. The root of a plant, like the mathematical root, is causative, the former being embedded in the earth, the latter embedded in the square. The visible growth of the plant, its proliferation into specificity, depends upon the root for stability and nutrition. The plant root nourishes because it is able to break down (divide) the fixed, dense mineral constituents of the soil into compounds which the plant can transform into its own tissue. In the vital sense the geometric root is an archetypal expression of the assimilative, generating, transformative function which is root. Like the vegetal root, the root of 2 contains the power of nature which destroys in order to progress (it severs the initial square) and it also contains the power which instantaneously transforms 1 into 2. The plant grows progressively out of a previous breaking down,

The ubiquitous 1 : √2 relationship is fundamental to the design of this Islamic floor mosaic as well as to the form and proportions of the honey bee.

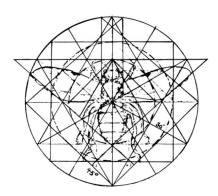

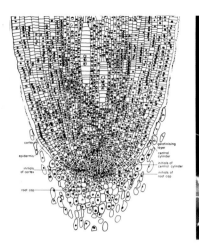

The morphic similarity between lightning and the root of a plant is also functionally accurate. Science now speculates that early in earth's evolution horrendous lightning storms in the atmosphere provided energetic ultra-violet light which transformed methane, hydrogen, nitrogen and carbonic gases into the proto-molecules for organic compounds. These molecules were deposited by torrential rains into the primal seas out of which life arose. Once again the 'root' functions as a transformative principle which supports the uplifting propensity which we call life.

The root grows by the constant division of its square shape. The root cells are a powerful metaphor for the principle of integration and transformation. Geometric contemplation is founded on the idea that natural forms are to be understood as symbols revealing metaphysical archetypal principles which guide and control universal evolution. The root contains an incredible power of growth; roots have been known to burrow over 100 feet through desert sand in order to reach water. A single tuft of rye grass may have over a billion root hairs, which laid end to end would extend for 350 miles. Roots aggressively hunt and battle in competition for water, air and minerals. They must constantly secrete acids to dissolve minerals to provide nourishment and protection for the plant. The root is a symbol for the law of sacrifice in nature, for, like a mother, its efforts are not for its own benefit but to uplift the plant in its movement toward light.

but there is no rational theory which can explain how a flower or a squash can spring forth from a tender, narrow stem, like the explosion of one square out of another. This is a transformative power existing *a priori* in the causal root.

The root principle is expressed in our bodies in the intestinal function, which is a transformation of food substance into energy. It is again expressed in the convolutions of the brain, which is related to the intestine in that it transforms crude, amorphous mental stuff into reason and understanding. The phallic or procreative power is implicit in the root, and the sexual function as well as the digestive function acts to root us in the physical world. We can see in the ancient agrarian practice of erecting stone monoliths, the phallic, mineral roots of the earth, the function of attracting downwards the fertile, cosmic ambiance. On the other hand, lightning is the root of the sky, transforming carbon and nitrogen into compounds assimilable for plants.

If we divide the full height of the human body into the harmonic divisions of the square root of 2, calling the total height unity, we locate the vital centre corresponding to what the Japanese call *hara* (belly), a subtle physical centre, just below the navel. The figure will measure $2-\sqrt{2}$ from the soles of the feet to the navel centre, and $\sqrt{2}-1$ from the navel centre to the top of the head. In Zen practice this centre is associated with a meditational technique of rooting, involving an intensification of the powers of physical self-control and self-transmutation. Tantric teaching in India, on the other hand, seeks to elevate this serpent or root so that it lends its energy to the higher, transformative glandular centres. The Chinese tradition speaks through Lao-Tzu, who said something like the following (my paraphrase):

Do not fear the ageing of the body, for it is the body's way of seeking the root. To seek the root is to return to the source, and to return to the source is to pursue one's destiny. And to pursue one's destiny is noble, and nobility is full of courage, and the courageous are those who seek to fulfil the spiritual goal behind all forms. So to seek the root is to pursue this goal.

The square root of two is the geometric function which represents the universal metaphor of the root, and the root represents the principle of transformation. This moment of transformation is everywhere before us, in the roots of plants transforming mineral into vegetal, in the leaves transforming sunlight into live supporting tissue, in rock and stone being weathered and worn down into molecular gases and liquid, liquid into gas, gas into solid, light into heat, heat into mechanical movement; in the germination of a seed. Shellfish convert phosphorus and sodium into their calcium shells★; the assimilation of food supports the creation of mental and spiritual experience. Everything is in a state of digestion, assimilation, transmutation. This transformation goes on in every passing moment as well as in the long

★ For a development of the theory of low-energy transmutation of elements in living systems see *Biological Transmutations* by Louis Kervran, Swan Books, 1976.

From one cell to two there is a cycle of change, in eight phases with seven intervals, analogous to the musical octave, or the spectrum of light. Seven symbolizes such cycles; the lunar month, a perfect example of graduated phases within a continuous process, is dominated by seven and its multiples. Seven relates more to process than to form, so there is no simple, natural way to draw a heptagon from a circle.

The functional pattern of the human nervous system is also sevenfold. From bottom of diagram: 1 Intrasegmental reflex: response limited to segment stimulated. 2 Intersegmental reflex: impulse carried by association neurones to neighbouring segments, causing co-ordinated muscle response. 3 Equilibratory control: automatic balancing reactions. 4 Synergic control: automatic co-ordinating control of muscular actions. 5 (a) Auditory and (b) visual reflexes: automatic responses to sudden noise or flashes of light. 6 Automatic associated control of complex muscular actions. 7 Voluntary and inhibitory control: choice of responses based on memory of past experiences.

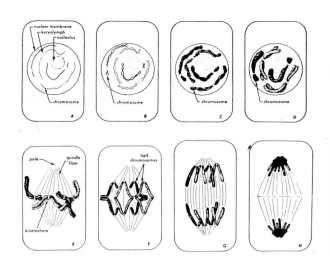

aeons of evolutionary cycles. Transformation is the ubiquitous condition of the worlds, and their evolution from mineral to plant to animal, kingdom emerging out of kingdom, volume forming itself out of the converging vector extensions of a preceding volume (see p. 72). There is periodicity, rhythm, oscillation, pattern, frequency, all measurable in time and space units. This is the genesis of sequential appearances, but the moment itself of transformation, from one state to another, from one quality of being to another, from one form or level of consciousness to another, is always a leap, a jump, an incomprehensible velocity, as it were, outside of time, as when one cell divides into two. If we approach life or evolution with only the sequential intelligence, only the rational, measuring facility, the reality of genesis will always elude us. This transformative moment is all that really exists; the phenomenal worlds are a transitory reflection. They are the past and future of this one ever-present eternity, the only possible eternity without duration which is the present moment.

In summarizing what we have observed in Workbook 1, let us philosophically raise square 1 to represent the principle of Unity, or that quality of absolute Unity which is represented in the finiteness of the square as a unit, an individual, a wholeness, or a system. Square 2 can be likewise extended to represent Duality, and the power of proliferation that is multiplicity. When one becomes two, we have automatically the potential of endless multiplicity through progression, as we have seen. Thus the extreme, essential polarity of the universe, Unity and Multiplicity, is perfectly represented and observable in the simple drawing of the square and its diagonal.

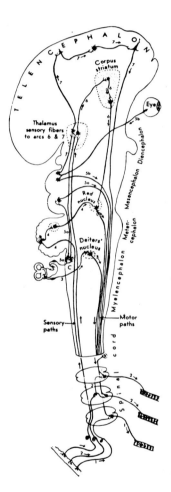

Let us now extend this envisioning of the simple, incommensurable root powers as geometric metaphors for the supra-rational moment of transformation to include not only the square root of 2 but also the square roots of 3 and 5, as has been done in all the known traditions of Sacred Geometry.

Transformation can be seen to occur by means of three general processes: the Generative, symbolized by the square root of 2; the Formative, symbolized by the square root of 3; and the Regenerative, symbolized by the square root of 5, and its related function of phi ϕ, the Golden Mean (to be discussed in Chapter V).

The square root of 3 appears in two major geometric configurations, and each of these demonstrates in a different way its *formative* character. The primary one, known as the Vesica Piscis (literally, a bladder [*vesica*] which when filled with air would be in the form of a fish [*piscis*]), was the central diagram of Sacred Geometry for the Christian mysticism of the Middle Ages. It is constructed by drawing two equal circles so that the centre of each lies on the circumference of the other. The second configuration in which $\sqrt{3}$ appears is that of the cube cut by its diagonal.

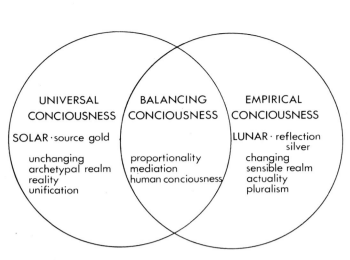

UNIVERSAL CONCIOUSNESS

SOLAR · source gold

unchanging
archetypal realm
reality
unification

BALANCING CONCIOUSNESS

proportionality
mediation
human conciousness

EMPIRICAL CONCIOUSNESS

LUNAR · reflection silver

changing
sensible realm
actuality
pluralism

One of the ways to view the Vesica Piscis is as a representation of the intermediate realm which partakes of both the unchanging and the changing principles, the eternal and the ephemeral. Human consciousness thus functions as the mediator, balancing the two complementary poles of consciousness.

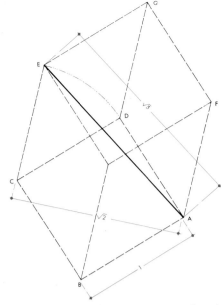

A cube with edges equal to 1; a rectangular plane is passed diagonally through the cube. Sides ED and $FB = 1$, and EF and $DF = \sqrt{2}$. Therefore the diagonal EB of both the plane and the cube equals $\sqrt{3}$.

Workbook 2
The $\sqrt{3}$ and the Vesica Piscis

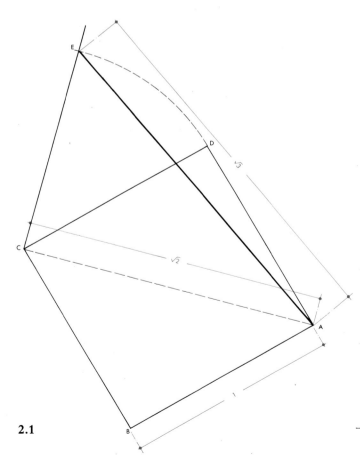

Drawing 2.1. Draw square $ABCD$ (it is shown here at a 30° angle so as to be visually comparable with the figure above). Draw diagonal CA. Draw a line perpendicular to CA at point C (see Drawing 1.1 for method). With centre C and radius CD equal to 1, swing an arc cutting this line at E. (Note that this operation is also illustrated by the dotted arc on side $EGCD$ of the cube above.)

As the division of Unity symbolized by the two-dimensional square yields the $\sqrt{2}$ function, so the division of Unity symbolized by the cube (representing three-dimensional volume) yields the $\sqrt{3}$ function.

Drawing 2.2. The construction of the Vesica Piscis. Draw a circle of any radius about any centre A. At any chosen point on the circumference of this circle, B, swing another circle of equal radius.

As the initial circle (Unity) projects itself outward in a perfect reflection of itself there is an area of overlap defined by the two centres (points A and B) and the intersection of the two circumferences. This area and shape is known as the Vesica Piscis.

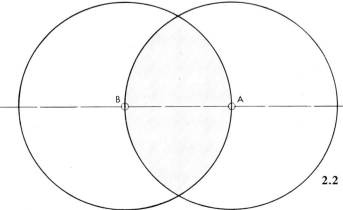

2.1

2.2

Drawing 2.3. Geometric proof of the $\sqrt{3}$ proportion within the Vesica Piscis. Draw the major and minor axes CD and AB. Draw CA, AD, DB and BC. By swinging arcs of our given radius from either centre A or B we trace along the vesica to points C and D, thus verifying that lines AB, BC, CA, BD and AD are equal to one another and to the radius common to both circles. We now have two identical equilateral triangles emerging from within the Vesica Piscis. Extend lines CA and CB to intersect circles A and B at points G and F. Lines CG and CF are diameters of the two circles and thus twice the length of any of the sides of the triangles ABC and ABD. Draw FG passing through point D.

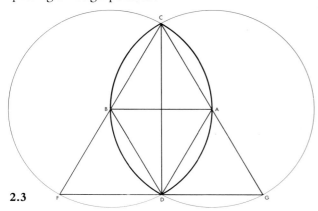

2.3

By the same method as above, we can prove that FD and GD are also equal to the sides of the triangles ABC and ABD.

If $AB = 1$, then $DG = 1$, $CG = 2$ and by the Pythagorean Theorem $(a^2 + b^2 = c^2)$ the major axis $CD = \sqrt{(CG^2 - DG^2)} = \sqrt{3}$.

Drawing 2.4. Geometric construction of the $\sqrt{3}$ Rectangle. From point O, the centre of the Vesica Piscis, draw a third circle with a radius of 1 and a horizontal axis bisecting all three circles and cutting the third circle at E and F. With points E

and F as centres, swing arcs with compass unchanged, cutting the new circle at points H, I, J and K. Draw the root 3 rectangle $HIJK$ enclosing the Vesica.

$$HI = OI = \text{radius } AB = 1$$
$$HK = CD = \sqrt{3}$$

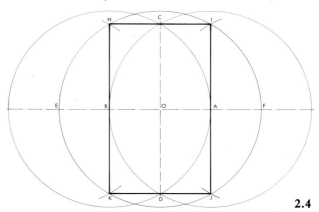

2.4

Drawing 2.5. Construction of the hexagon from the Vesica Piscis. With our Vesica $ABCD$, swing an arc from C as centre and original radius $1 = CB$, cutting the second circle at E. Repeat with D as centre cutting the circle at G. Repeat again with either E or G as centre cutting the circle at F. Draw hexagon $BCEFGD$.

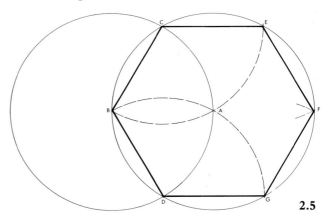

2.5

There are few figures which carry so much meaning as the simple Vesica. Keith Critchlow has explored this form with great depth and sensitivity in his book *Time Stands Still*, and in the exposition of the geometry of Chartres Cathedral in his beautiful film, *Reflections*, so I will consider here only a few of its symbolic interpretations.

The overlapping circles – an excellent representation of a cell, or any unity in the midst of becoming dual – form a fish-shaped central area which is one source of the symbolic reference to Christ as a fish. Christ, as a universal function, is symbolically this region which joins together heaven and earth, above and below, creator and creation. The fish is also the symbolic designation of the Piscean Age, and consequently the Vesica is the dominant geometric figure for this period of cosmic and

Commentary on Workbook 2

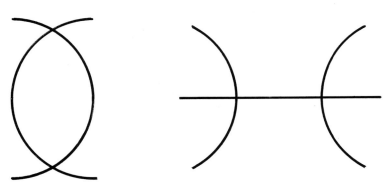

The variations on the symbol for the zodiacal sign Pisces
relate it to the vesica.

The succession of polygons as they arise out of the scission of unity.
As unity, represented by the circle, self-divides, its centre becomes the
duality of points A and B. The line AB naturally unfolds in the equilateral triangle (thus, all things being dual by nature are 3 by principle). As the equilateral triangle unfolds outwards it successively
defines the sides of the square (4), the pentagon (5), the hexagon (6),
the octagon (8), the decagon (10) and the dodecagon (12).
   To construct this figure, draw the originating circles forming the
Vesica Piscis, then draw the additional circles as shown. Various points
of intersection when projected will define the vertices of the various
polygons (shown in coloured dotted lines). The black dotted lines
indicate further points of concordance and help define more vertices.
The coloured dotted lines indicate the location of the pentagon, since
this is not an obvious connection of points (see Workbook Drawing
2.6).
   This drawing of growth even suggests the tree. The vesica can represent the seed. From its germination spring forth the coloured circles
(the root) and the polygons (the germ, giving rise to the branches).
The $\sqrt{3}$ contained within the Vesica Piscis is the formative power
giving rise to the polygonal 'world'.

Christ within the vesica.

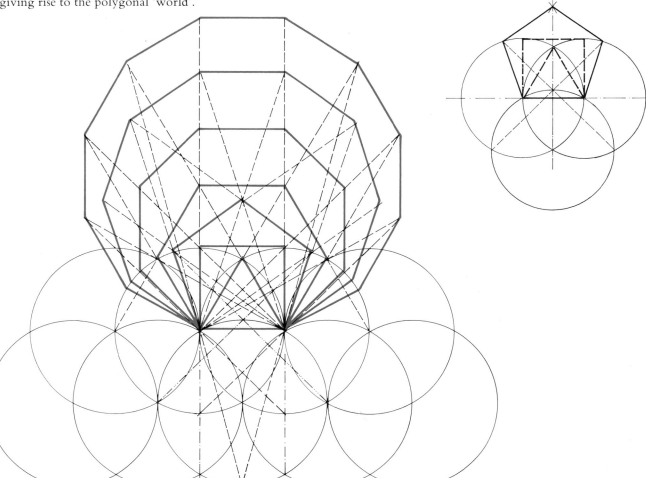

human evolution, and is the major thematic source for the cosmic temples of this age in the west, the Gothic cathedrals.

Jesus as the centre of the Vesica carries the idea of the non-substantial, universal 'Christic' principle entering into the manifest world of duality and form. The Piscean Age has been characterized as that of the formal embodiment of spirit, manifesting a deeper penetration of spirit into form, with a concurrent deepening of the materialization of spirit: the Word becomes flesh. Thus the square root of 3 is linked to the formative process, and this connection is further clarified when one observes the relationship of the Vesica and the square root of 3 to the hexagon, which is the symmetry of order for the measure of the earth, the measure of time (through the 360° of the Great Circle of the heavens), and also the basic formation of mineral crystals, especially of the carbon bonding patterns which allow for the formation of all organic substances. To regard this principle of formation from a more strictly geometric point of view, we find that while $\sqrt{2}$ divides the surface of the square, the $\sqrt{3}$ divides the volume-form of the cube, and we should recall that everything in the created universe is a volume. The formation of any volume structurally requires triangulation, hence the trinity is the creative basis of all form. The cube is the most elementary symbol of the manifest (volumized) formal world.

The Vesica is also a form generator in that all the regular polygons can be said to arise from the succession of vesica constructions.

The roots of 2 and 5 can also be derived from this cosmogram of the Vesica, because there is no synthetic symbolization of Unity which does not evoke all the major principles (see p. 37): as the *Koran* says, 'There is no god which is not all gods'. But the Vesica emphasizes the $\sqrt{3}$ with the rich texture of contemplation that this symbol evokes.

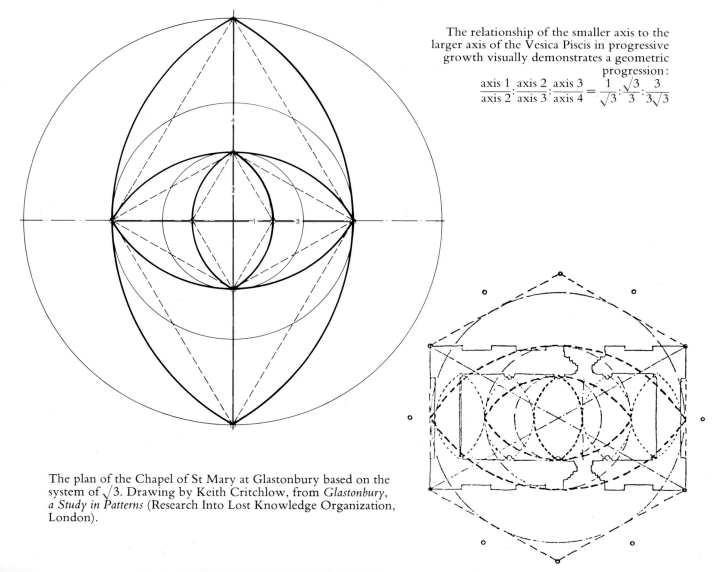

The relationship of the smaller axis to the larger axis of the Vesica Piscis in progressive growth visually demonstrates a geometric progression:

$$\frac{\text{axis 1}}{\text{axis 2}} : \frac{\text{axis 2}}{\text{axis 3}} : \frac{\text{axis 3}}{\text{axis 4}} = \frac{1}{\sqrt{3}} : \frac{\sqrt{3}}{3} : \frac{3}{3\sqrt{3}}$$

The plan of the Chapel of St Mary at Glastonbury based on the system of $\sqrt{3}$. Drawing by Keith Critchlow, from *Glastonbury, a Study in Patterns* (Research Into Lost Knowledge Organization, London).

Workbook 3
The $\sqrt{5}$

Drawing 3.1. Generating the $\sqrt{5}$ rectangle from the 1:2 rectangle. Starting with a double square *ABCD*, bisected at *EF*: with centre *G* and radius *GA*, swing a semicircular arc intersecting the extended dividing line *EF* at *H* and *K*.

$HK = \sqrt{5}$. *MLKH* is a $\sqrt{5}$ rectangle.

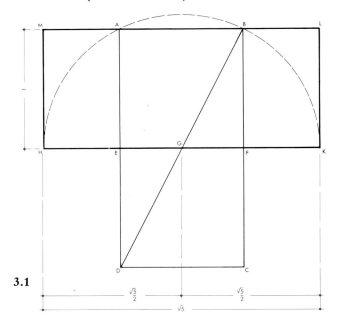

3.1

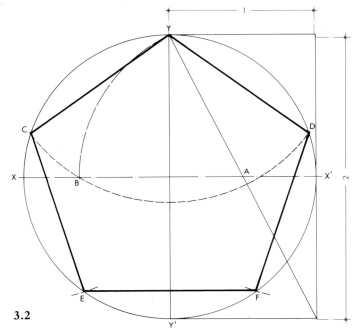

3.2

Drawing 3.2. The $\sqrt{5}$ and the pentagon. Draw a circle, with its semicircle inscribed in a double square rectangle as shown. Extend the dividing line of the double square to complete the two cardinal axes, *XX'* and *YY'* of the circle. With centre *A* and radius $AY(=\sqrt{5}/2)$ swing an arc to *B*. With centre *Y* and radius *YB* swing an arc to cut the circle at *C* and *D*. With centres *C* and *D* and compass unchanged swing two more arcs cutting the circle at *E* and *F*. Draw pentagon *YDFEC*.

These geometric demonstrations reveal the relationship of $\sqrt{5}$ both with the number 5 (as the square of $\sqrt{5}$) and with the fivefold symmetry of the pentagon.

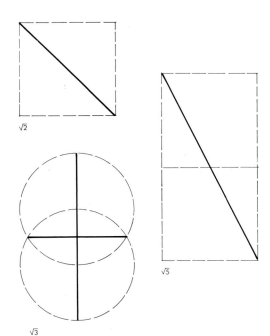

The appearance of the three sacred roots can be summarized in this simple diagram. These three root relationships are all that are necessary for the formation of the five regular ('Platonic') solids which are the basis for all volumetric forms. Also, 2, 3, and 5 are the only numbers required for the division of the octave into musical scales. We can then accept these roots as a trinity of generative principles.

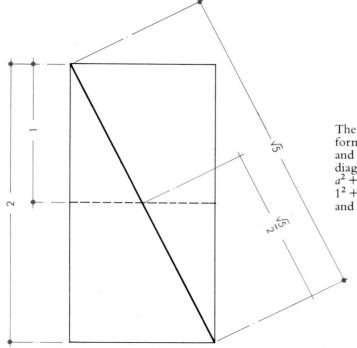

The double square divided by a single diagonal forms two right triangles, each having a base of 1 and a height of 2. To find the geometric value of the diagonal we apply the Pythagorean formula $a^2 + b^2 = c^2$. In this case, $a = 1$, $b = 2$, therefore $1^2 + 2^2 = c^2$ or $1 + 4 = 5$, so that the diagonal $= \sqrt{5}$, and the semi-diagonal of a single square $= \sqrt{5}/2$.

It seems that the dividing and transforming root powers must be seen at the same time as powers that bind and synthesize, as such principles must often demonstrate two poles of an opposition. The square root of 5 transverses two worlds, indicated by the upper and lower squares, the world of spirit and the world of body. And all the forms of bonding or the mediating principles between these cosmic extremes we will consider as the 'Christic Principle'. The $\sqrt{5}$ is the proportion which opens the way for the family of relationships called the Golden Proportion. The Golden Proportion generates a set of symbols which were used by the Platonic philosophers as a support for the ideal of divine or universal love. It is through the Golden Division that we can contemplate the fact that the Creator planted a regenerative seed which will lift the mortal realms of duality and confusion back towards the image of God.

We shall examine the Golden Section and its ramifications shortly. But first let us look at the principle governing the progressions which result from the sacred roots of 2, 3, and 5.

Commentary on Workbook 3

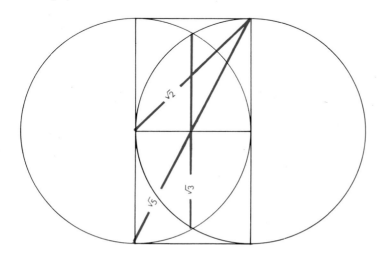

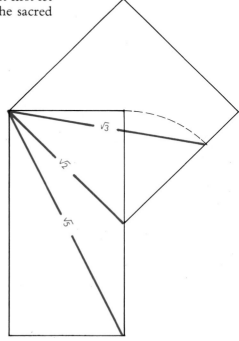

The two principal elements of sacred geometry, circle and square, in the act of self-division give rise to the three sacred roots. The roots are considered as generative powers, or dynamic principles through which forms appear and change into other forms.

IV Alternation

We have emphasized the fixed, invariable quality of the incommensurable root relationships to Unity as they appear in geometric figures. This is analogous to the stabilizing role that the root function plays in the growth of a plant. But the root also is the generator of change in the continuum of the ever-moving, irreversible phases which are a part of organic life.

Because the ancients thought as geometers there was for them no separation between geometry and natural science or cosmology or theology. The conformity of mathematics to the natural laws of geometry led directly to one of the major philosophical premises of ancient thought, that of *alternation*. In this chapter we will examine how ancient methods of calculation reveal and rely upon this universal law.

Ancient mathematics had no decimal system with which the numerical equivalency of the incommensurable square root of 2 (1·4142135 . . .) could be indicated. This was more than a limitation in their notational system; the idea of an irrational number such as this was to the ancient geometer a logical absurdity. To him the essence of number was a state: tangible, fixed, measurable. *Ratio*, the Latin root of 'reason', also means 'measure'; an irrational number was an unacceptable contradiction.

The two types of numbers, rational and irrational, represented two completely different states of being. The whole numbers were related to manifestation and were the terms to be used in calculation. Every aspect of the phenomenal world was seen to be a fixed, instantaneous moment caused by the interaction of complementary components, a moment trapped between light and dark, life and death, day and night, between formation, disintegration and reformation. An arrested formation was epitomized in ancient geometry by the Diaphantine triangle, which is a right-angled triangle with all three sides equal to whole numbers, such as the 3, 4, 5 triangle. This latter is traditionally called the Sacred Triangle, 'sacred' meaning fixed or permanent, thus symbolically related to the fused sacral bones of the spine which make possible the stable, seated posture.

On the other hand, the irrational roots symbolize the constant, creative process of acting and reacting energy. This immeasurable gestating force emanates from the incomprehensible Unity. That which is comprehensible is no other than a momentary limitation of this One, indefinable Being into a definable moment: 'Necessarily, then, all that is definable arises out of an Indefinable All.'

Derivation of the sacred 3, 4, 5 triangle by the crossing of three semi-diagonals ($\sqrt{5}/2$), showing also its geometric proof. This diagram demonstrates the relationship in Sacred Geometry between process and structure. The irrational roots, such as the $\sqrt{5}$, are symbols of pure archetypal processes (generation, fusion, transformation, etc.), while the fixed whole number relationships are the structures which emerge to symbolize these process principles. In this figure the crossing of two irrational lines ($\sqrt{5}/2$) produces the 3, 4, 5 'Pythagorean' Triangle, the figure upon which rests the rationality of our mathematical thought.

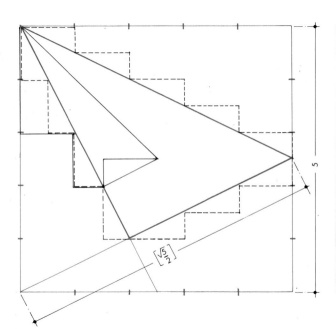

Design of a page from the Lindisfarne Gospels (*c.* AD 700) with proportions based on the 3, 4, 5 triangle.

But the reverence which shaped the ancient mathematician's thought did not preclude the use of these principles in calculation. In several pre-Euclidian mathematical texts a method is given which allows for the expression of these root powers as a succession of whole number ratios. These ratios emerge in such a way that they go alternately above and below the incommensurable root value. In addition to the alternating pattern, these successive ratios approach nearer and nearer to the root value with each alternation. Expressed in this way, the roots retain their dynamic or 'process' quality and at the same time reveal the Principle of Alternation.

Theon of Smyrna, a second-century AD Platonist philosopher and mathematician, in his book *The Mathematics Useful for Understanding Plato*, transmitted a demonstration of what are called the lateral and diagonal numbers. I will include here Theon's complete rhetoric on this problem, which on first reading will seem like a non-sense puzzle. Yet if one follows the numerical and geometric procedure the confusion will disappear and at the same time the technique of calculation will become clear together with its philosophical implications.

Theon begins this demonstration with a square as unity, which he declares to have both a side and a diagonal equal to the value of 1. This description indicates an esoteric significance, because the square with side and diagonal both equal to 1 is to our mentality an absurdity. But it conforms to a mystic fidelity to the sense of Unity held by the ancients, for which all aspects or differentiations, be they the principle of the side of the square or its diagonal, are as one and equal to one when contained in the original unity. We shall see when we come to discuss the spirals that other numerical progressions also necessarily begin with this double 1; its usefulness will become apparent, if for a moment we follow Theon and use it.

Here is Theon's demonstration, which will be followed by the same concept rendered geometrically:

> Just as the numbers have in power relationships with triangles, tetragons, pentagons and other figures, so also we find that the relationships of lateral numbers and diagonal numbers are manifested in numbers according to the generative ratios, because these are the numbers which harmonize the figures. Therefore, as Unity is the principle of all figures, according to the supreme generative ratio [that is the ratio of 1 to 2], likewise the relationship of the diagonal to the side finds itself within Unity. Let us suppose, for example, two unities, one of which is the side and the other the diagonal, for it is necessary that the Unity, which is the principle of all, be in power both the diagonal and the side. Let us add to the side the diagonal, and to the diagonal, let us add two sides, because what the side can do two times the diagonal can do once.

This simply means that twice the square of the side equals the square of the diagonal. He continues,

> From then on the diagonal becomes larger than the side, for in the first side and the first diagonal, the square of the unity-diagonal will be one unity less than the double square of the unity-side, because the unities are in equal equality, but one is less by one unity than the double of the unity. Let us now add the diagonal to the side, that is to say unity to unity, and the side will have the value of two unities; but if we add two sides to the diagonal, that is to say, two unities to the unity, the diagonal will have the value of three unities. The square constructed on side 2 is 4, the square of the diagonal is 9, which is one unit larger than the double of the square of 2.

> In the same way, let us add to side 2, diagonal 3. The side becomes 5. If to the diagonal 3 we add 2 sides, that is 2 times 2, we will then have 7 unities. The square constructed on the side 5 is 25, and the one constructed on the diagonal 7 is 49, which is one unity less than the double of the square of 25. Again, if to the side

5 one adds the diagonal 7, 12 unities are obtained, and if to the diagonal 7 one adds two times the side 5, 17 unities are obtained, whose square is 289, which is one unity larger than the double square of 12 (288), and continuing thus, the proportions alternate; the squares constructed on the diagonal will be sometimes smaller, and sometimes larger by one unity than the double of the square constructed on the side, such that these diagonals and these sides will always be expressible.

Workbook 4
Alternation

EXPLANATION OF THEON'S DEMONSTRATION

We begin this demonstration with a theoretical relationship between a square and its diagonal, the Unity (original square, the monad) with both side and diagonal valued at 1. We continue to generate theoretical side to diagonal relationships following the pattern (given by Theon) of adding the diagonal and side of square 1 in order to produce the side of square 2 and adding double the side of the first square to the diagonal of square 1 to obtain the diagonal of square 2. The initial step and the procedure may at this point sound absurd, but accept it for the moment and you will see how it works geometrically.

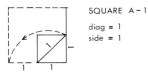

SQUARE A-1
diag = 1
side = 1

Add the value of the diagonal of square 1 to the side of Square 1 in order to obtain the side of square 2: $1 + 1 = 2$.

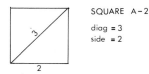

SQUARE A-2
diag = 3
side = 2

Add the double of the side of square 1 to the diagonal of 1 to obtain the diagonal of square 2, that is $1 + 2 = 3$.

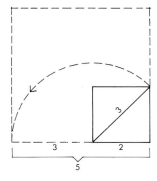

Then add the value of the diagonal of square 2 to the side of square 2 in order to obtain the side of square 3: $2 + 3 = 5$.

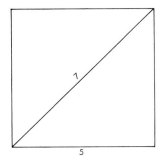

SQUARE A-3
diag = 7
side = 5

Next add the double of the side of square 2 to the diagonal of square 2 to obtain the diagonal of square 3: $3 + (2 \times 2) = 7$.

The side to diagonal relationship of the theoretical squares changes from $1:1$ to $3:2$ to $7:5$. Square 4 will have a diagonal of $7 + (2 \times 5) = 17$ and a side of $5 + 7 = 12$.

To continue this generation we follow the same rule: add the value of the side of the square to the value of the diagonal which gives us the value of the side of the next larger square, and then add *double* the value of the side to the value of the diagonal to give the value of the diagonal of the next larger square:

Square	1	2	3	4	5	6
side	1	2	5	12	29	70
diag. (root)	1	3	7	17	41	99

The root to side relationships, $2:3$, $5:7$, $12:17$, $29:41$, etc. provide coefficients which by the fifth expansion have produced a highly accurate decimal equivalency to our presently utilized $\sqrt{2}$ ($41/29 = 1.414286\ldots$). These coefficients oscillate first above then below then again above, coming closer and closer to the perfect irrational state. This clearly expresses, in addition to rhythmic alternation, the concept of a movement towards perfection as the manifesting aspects of growth draw nearer and nearer to the causative root power. The power of scission contains within itself the power of the return to the cause.

Lateral number	Square	Double square	Diagonal number	Square on diagonal number	Difference
1	1	2	1	1	2−1
2	4	8	3	9	8+1
5	25	50	7	49	50−1
12	144	288	17	289	288+1
29	841	1682	41	1681	1682−1

This progression can be continued indefinitely, and the above table verifies Theon's mysterious assertion that the square on the diagonal will always be double the square of the side, but alternatively larger or smaller by one unity.

Drawings 4.1 and 4.2. The theoretical numerical progression of side to diagonal ratios is placed next to the geometric development to show graphically how quickly the whole number sequence approaches the irrational $\sqrt{2}$ function. From the given unity square with A as centre and AA' as radius, swing an arc cutting the X axis at B. With Y as centre and radius YB, swing a semicircle cutting the Y axis at B'. With B as centre and radius BB' swing an arc cutting the X axis at point C (5 units). With Y as centre and radius YC swing a semicircle cutting the Y axis at C' to determine square 3 and its germ along the X axis. Repeat to draw squares 4, 5 . . .

The root of square 1 becomes the germ of 2; the root of square 2 becomes the germ of 5; the root of square 5 becomes the germ of 12.

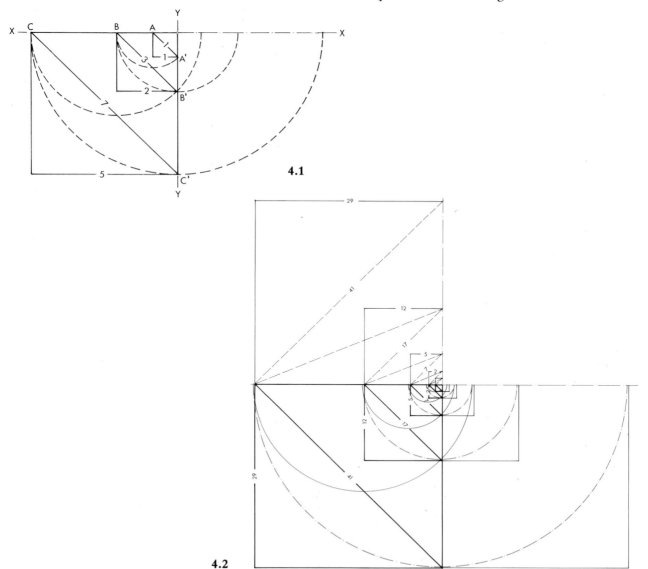

4.1

4.2

Drawing 4.2, based on Theon's demonstration, is from *Le Temple de l'Homme* by R.A. Schwaller de Lubicz, and it presents a growth pattern through the root of 2 by which all of nature proceeds. What is revealed here is a precise demonstration through the root of 2 of the Principle of Alternation, an alternation both in power – the energetic, causal pulsation of the supra-rational root – and also in the formal oscillation of squares produced by this power.

When we look back at our table of root to side relationships, 3 to 2, 7 to 5, 17 to 12, 41 to 29, we see that coefficients are provided which, by the fifth or sixth expansion, have produced a ratio equal in accuracy to the presently used square root of 2, and that starting the progression with side and diagonal equal was functionally accurate. Each coefficient oscillates first above then below, coming closer and closer to the perfect irrational state. This is a basic element in what is called Diaphantine Mathematics, which sets up numerical progressions which can be seen as representations of vibrational systems in that a vibrating string also moves above and below an abstract node or inexpressible still point. We may more poetically conceive of this as a model of the pulsation of Cosmic Life.

The Principle of Alternation has been a source of metaphysical and physical wisdom in many great cultures of the past. Today we have become most familiar with it in the Taoist philosophy through the widespread study of Zen Buddhism, which owes much to it, and of the I Ching.

To the Pythagorean demonstration can be added R.A. Schwaller de Lubicz's marvellous insight of the *germ*. When the root, with its power of multiplicity and growth and proliferation, is projected outside of the unity it forms in relationship to 2 a remaining segment which, geometrically, behaves similarly to the plant germ. I'm referring here to the root principle which contains a power which botanists call 'positive geotropism', in other words, the power to descend, involve and transmute from below. The germ, then, represents the power of 'negative geotropism', or that which causes growth upwards and outwards – in other words the entire ascent culminating in the new seed. These then are polarized, directional opposites of the same power. If a seed is placed upside down the root will immediately direct itself downwards, while the germ carrying the stem will make the turn and grow upwards. A Taoist teacher would say of this that all of life and the entire universe progresses through alternation. The truth of every progression or evolution is rhythmic alternation and oscillation. Everything alternates towards its opposite. As far as natural and cosmic movement is concerned, the only inevitability is alternation.

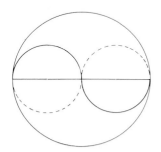

If we make a diagram of the pattern of Theon's progression, which alternates above and below, yet continuously nearer to, an irrational centre, we will have a general convergence wave pattern. Computer analysis shows that these ratios, after many alternations, come to an extremely close approximation to the irrational root, then gradually move away. We have, then, a general convergence–divergence configuration. The curve can also be drawn to indicate three dimensions, giving us the image of a spiral with its mirror reflection, the Taoist image for the movement of the great cycles of time.

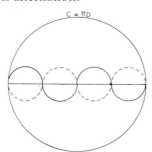
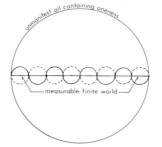
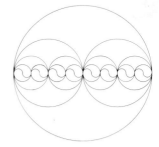

The principle of alternation is expressed geometrically in the ancient Taoist symbol of yin and yang. The form of this symbol arises from two equal circles within the larger circle, the diameter of each smaller circle being exactly 1/2 that of the larger. The ratio of the diameter to the circumference of any circle is π; $C/D = \pi$.

At first glance the symbol suggests that the division of Unity (here taken as the large containing circle) is into two equal parts. Such a division results in a static equilibrium, without any possibility for growth. It is the asymmetrical division, as already demonstrated in the $1:\sqrt{2}$ relationship, which creates proportion and thus the progression into form which we call growth. Later, in the Squaring of the Circle, we will discover the asymmetrical principle hidden within this symbol. But it is important to note in this context that the circum-

ference of the smaller circles equals $D/2 \times \pi = \pi D/2$. The sum of the circumferences of the two inner circles equals the circumference of the large circle ($2 \times \pi D/2 = \pi D$). The figures show the continuation of this initial division carried through into 4 and then 8 divisions. This process of halving the circles can be carried on indefinitely; and at any point, when the sum of the circumferences of the smaller circles is totalled, it will still equal the original larger circle. This process can be taken to the point where the wavy line and the diameter become indistinguishable from one another, illustrating the paradox of the diameter becoming equal to the circumference of the same circle. Thus, like Theon's demonstration, this ancient diagram shows that at the origin and end all differentiation merges towards Unity.

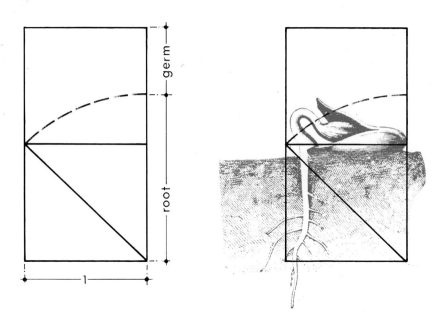

The universal dichotomy is expressed in every germinating seed. The seed immediately divides into root and germ. There is an *alternation of function*, as the germ provides the nourishment until the root begins to function, then the germ transforms into the first leaves, leaving behind the seed shell, and the root takes up the nutritive work. This alternating function of root/germ is symbolized geometrically in Workbook 4 (drawing 4.2), where the root of one square is equal to the germ of the next square and so on in each successive square.

This figure illustrates a comparison, which, like all comparisons in geometric philosophy, is of a three-term proportional type: $a:b::b:c$. In this case, the geometric root/germ is related to the universal root/germ principle in the same way as this principle is related to the botanical expression of root and germ. We are geometrically exploring an analogical, proportional thought, rather than following the more rigid, equational logic.

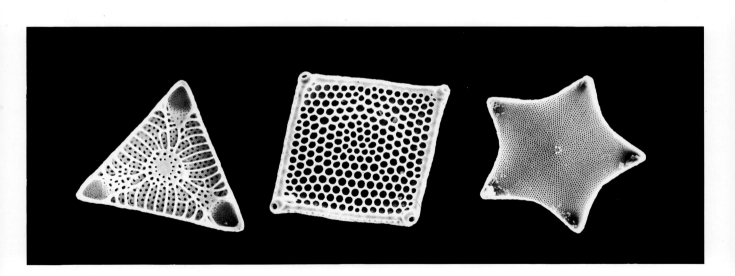

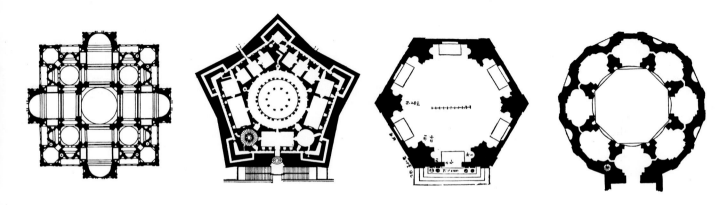

The numbers which emerge from the 3, 4, 5 'Pythagorean' triangle provide beautiful symmetries for natural forms. This series begins with a natural expression of the equilateral triangle and concludes with a series of symmetries used as the inspiration for ground plans in Renaissance architecture.

V Proportion and the Golden Section

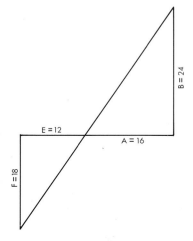

A four-term discontinuous proportion can be represented graphically by similar triangles positioned on a crossing of a horizontal and diagonal axis. To illustrate the proportion

$A : B :: E : F$

or $16:24::12:18 = 2/3$

draw line segment $E = 12$ and line $A = 16$ on the same horizontal with ends tangent at O. Raise a perpendicular B from the end of line A to establish any desired proportional relationship to 16, in this case, $B = 24$. The ratio $A:B = 2/3$. Draw a diagonal from the upper end of B passing through O. This diagonal will always intersect with a perpendicular dropped from the end of E so that line segment F will be in relation to E in the same ratio as B is to A, thus geometrically verifying that when one has three terms of a four-term proportion it is always possible to find the fourth term.

It was the goal of many traditional esoteric teachings to lead the mind back toward the sense of Oneness through a succession of proportional relationships. A proportion is formed from ratios, and a ratio is a comparison of two different sizes, quantities, qualities or ideas, and is expressed by the formula $a:b$. A ratio then constitutes a measure of a difference, a difference to which at least one of our sensory faculties can respond. The perceived world is then made up of intricate woven patterns of, as Gregory Bateson says, 'differences which make a difference'. Not only then is a ratio $a:b$ the fundamental notion for all activities of perception, but it signals one of the most basic processes of intelligence in that it symbolizes a comparison between two things, and is thus the elementary basis for conceptual judgement.

A proportion, however, is more complex, for it is a relationship of equivalency between two ratios, that is to say, one element is to a second element as a third element is to a fourth: a is to b as c is to d, or $a:b::c:d$. It represents a level of intelligence more subtle and profound than the direct response to a simple difference which is the ratio, and it was known in Greek thought as *analogy*.

When we think with four elements, that is with two different ratios, we locate our thought within the range of manifestation, of the natural world, since four is the number-symbol indicative of the finite, rational, measurable world of procreated form.

So, $a:b::c:d$ is a general formula of four related elements. The same thing is said numerically by $2:4::3:6$. The Pythagoreans called this procedure of thinking a *discontinuous proportion* of four terms.

When we further limit ourselves to three terms, that is when we lift ourselves one level to the realm of principles or activities (threeness), we find that the determination becomes more exacting through the reduction of the number of elements involved. Thus, one element is to a second element as the second element is to a third: $a:b::b:c$. Here the extremes are bound together by a mean term, b. The Greeks called this a *continuous proportion* of three terms, and this indicates a decided shift in the symbolization of perceptual and conceptual processes. Nicomachus and other Greek philosophers considered it as the only one which can be regarded as strictly *analogos*. It is the perceiver himself (b) who forms the equivalency or identity between observed differences (a and c). The perceiver no longer stands outside the comparative activity as in the four-term, discontinuous or disjunctive mode, which images the perceived difference as being separate ratios or distinctions.

Perhaps an example will be helpful here. Our experience of the world is due to our organs of perception being sensitive to variations of the wave frequency patterns which surround and pervade our field of awareness. We distinguish a red cup from a green tablecloth only because our optic nerves set up a brain wave pattern which corresponds to the frequency patterns emanating from the cup and tablecloth. The perceiver himself is then the indispensable bond in the registration of these variations in external frequency patterns, interpreting and distinguishing them as objects such as the cup and the tablecloth.

Many philosophers speak of reaching a state of consciousness in which one is constantly aware of this integration and attunement between the apparent external vibratory field and the inner field of perception. This mode of perceptual awareness, which we find comparable to a three-term, continuous proportion, was referred to by Sri Aurobindo as 'knowledge by identity', and regarded as an important stage in the process of spiritual development: while acknowledging an external source of experience we recognize that it is in a continual flow of relationship with our internal faculties of perception and cognition, and it is this relationship, not the external object itself, that we are experiencing. The objective world then is interdependent with the entire physical, mental and psychological condition of the perceiving individual, and consequently will be altered by changes in his inward condition. It is possible to become aware of drawing the external object out of the totality of our inner space, thus fusing together the contemplation of self and world.

Is, then, a three-term proportion as close as we can approach the sense of unity with proportional thinking? The response to this question is no, for there is one, and only one, proportional division which is possible with *two terms*. This occurs when *the smaller term is to the larger term in the same way as the larger term is to the smaller plus the larger*. It is written $a:b::b:(a+b)$. The largest term $(a+b)$ must be a wholeness or unit composed of the sum of the other two terms.

Historically this unique geometric proportion of two terms has been given the name 'Golden Proportion', and is designated by the 21st letter of the Greek alphabet, phi (ϕ), although it was known by cultures much older than the Greek.

There are two importantly different ways to consider this primary geometric proportion in relation to Unity. The first occurs when the largest term (in this case $(a+b)$ is greater than 1 or unity. The second case occurs when the largest term $(a+b)$ equals unity or 1 (in formula $a:b::b:1$). Each of these gives rise to an important characteristic of ϕ.

What we are following in this chapter is essentially a set theory description of all the possible types of geometric proportions. We have first isolated two major sets of geometric proportions, the four-term and the three-term. Within the three-term continuous proportion we have defined a special sub-set in which the third term is equal to the first term plus the second term, $a:b::b:(a+b)$, so that actually only two terms, a and b, are found in the three term proportion. This is called ϕ, the Golden Proportion. The fact that it is a three-term proportion constructed from two terms is its first distinguishing characteristic, and is parallel with the first mystery of the Holy Trinity: the Three that are Two.

In the first figure, two equal line lengths have been divided in such a way that $a:b::b:a+b$, or $b/a = \phi$. The first case shows a proportion in which the whole line is larger than Unity. Unity is defined as the segment b with the segment a, an expansion of it, attached to it making the whole line $a+b$. In proportional thought there are no fixed quantities, only fixed relationships. The quantitative value may shift but the relational configuration remains the same. Here we give $b=1$ to assure that the whole is more than unity, and is also a relational expansion of unity.

$$\text{First term} = a$$
$$\text{Second term} = b = 1$$
$$\text{Third term} = b+a = a+1$$

$$\frac{a}{b}:\frac{a}{a+b} \quad \text{or} \quad \frac{b}{a+1}$$

There are numerous examples of this type of proportion, where the third term $(a+b)$ is greater than one, in the ϕ progression as in the fundamental $\sqrt{2}$ proportion:

$$\frac{a}{b}:\frac{b}{a+1} = \frac{1}{\sqrt{2}}:\frac{\sqrt{2}}{1+1} \quad \text{or} \quad \frac{a}{b}:\frac{b}{a+1} = \frac{1/\phi}{1}:\frac{1}{1/\phi+1}$$

These two examples are drawn from families of three-term geometric proportions in which the third term is a relational expansion of unity, and therefore larger than unity.

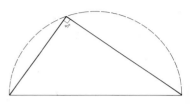

In order to represent a three-term, continuous proportion geometrically we can use the theorem of Thales which states that any angle inscribed in a semicircle will be a right angle.

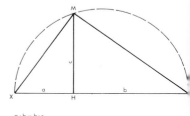

$a:b::b:c$

Draw line xy and from its centre at O draw a semicircle with xy becoming the diameter of the semicircle. Raise *any* line HM perpendicular to xy, terminating at the circumference. Connect points Mx and points My to form a right triangle xMy. We have then

$$\triangle xMy \approx \triangle xHM$$
$$\triangle xMy \approx \triangle MHy$$
$$\triangle xHm \approx \triangle MHy$$

By the law of similar triangles we find that the perpendicular MH is the geometric mean term between line xH and line Hy. Therefore the three-line segments will be the geometric representation of a three-term continuous proportion of the type

$a:b::b:c$.

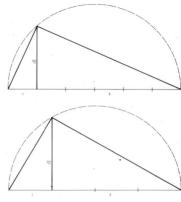

No matter where this perpendicular is raised on the diameter, it will always be the geometric mean term between the two segments of the diameter.

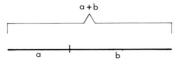

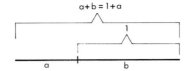

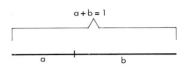

The line constitutes a wholeness, a unity.

First case: the whole is more than one.

Second case: the whole is equal to one.

In the second figure we shift the value of unity from the part to the whole, so that its divisions must be less than 1. In doing so we will find the second and completely unique characteristic of ϕ as the only geometric partitioning of Unity. This manoeuvre of shifting value is typical of numerous problems found in the oldest known mathematical texts from both Egypt and Babylonia, and was a basic technique of ancient mathematical procedure. In this case,

$$\text{First term} = a = 1 - b$$
$$\text{Second term} = b = 1 - a$$
$$\text{Third term} = a + b = 1$$

so, $\dfrac{a}{b} : \dfrac{b}{1}$

$$b^2 = a \times 1$$
$$b^2 = a$$
$$b = \sqrt{a}$$

This algebraic formulation is fully demonstrated geometrically in Workbook 5. Here we have the root of a as equal to the root of b^2, so that a and b are in relationship to one another as a root is to a square. This necessitates that as the third term in the geometric proportion, $a + b = 1$ must in this case be a square plus its root = 1. ϕ *is the unique division which fulfills this characteristic:* $1/\phi + 1/\phi^2 = 1$. This completes the mathematical metaphor for the Trinity: 'Three that are Two that are One'. It is the ultimate reduction of proportional thought to the causal singularity.

If we once again use proportion as a model for the perceptual activity based on the recognition of differences, we have in this unique Golden Proportion 'within' Unity a case where the perceived difference (that which we experience as an object) plus the perceiver of that object are symbolized as contained within a sustained awareness of an all-encompassing Unity, $a : b :: b : 1$. This perceptual state corresponds to the goal of dynamic meditation.

The Golden Proportion is a constant ratio derived from a *geometric* relationship which, like π and other constants of this type, is 'irrational' in numerical terms. I have therefore tried to avoid initially presenting the Golden Proportion as a numerical quantity, i.e. $\phi = 1 \cdot 6180339 \ldots$, or $\phi = (\sqrt{5} + 1)/2$, but instead have demonstrated that it is first and foremost a proportion, not a number, a proportion upon which the experience of knowledge (*logos*) is founded.

In a sense, the Golden Proportion can be considered as supra-rational or transcendent. It is actually the first issue of Oneness, the only possible creative duality within Unity. It is the most intimate relationship, one might say, that proportional existence – the universe – can have with Unity, the primal or first division of One. For this reason the ancients called it 'golden', the perfect division, and the Christians have related this proportional symbol to the Son of God.

Why, it may be asked, cannot Unity simply divide into two equal parts? Why not have a proportion of one term, $a : a$? The answer is simply that with equality there is no *difference*, and without difference there is no perceptual universe, for, as the Upanishad says, 'Whether we know it or not, all things take on their existence from that which perceives them.' In a static, equational statement one part nullifies

46

the other. An asymmetrical division is needed in order to create the dynamics necessary for progression and extension from the Unity. Therefore the ϕ proportion is the perfect division of unity: it is creative, yet the entire proportional universe that results from it relates back to it and is literally contained within it, since no term of the original division steps, as it were, outside of a direct rapport with the initial division of Unity. This is the essential difference between the division of Unity by the square root of 2 and its division by ϕ, both of which are geometrical proportions. As the geometry of the former shows, through the creation of $\sqrt{2}$ we are immediately propelled outside the original square (see Workbook 1). This marks the beginning of an endless, ever expanding progression and proliferation, leading further and further away from the original Unity. There is no possible way with $\sqrt{2}$ to have a geometric *internal* division of Unity. The division by ϕ, on the other hand, gives a model of evolution which has as its goal the image of the perfection of the original Unity.

Progression by the Golden Division
$$\frac{1}{\phi^3}:\frac{1}{\phi^2}::\frac{1}{\phi^2}:\frac{1}{\phi}::\frac{1}{\phi}:1::1:\phi::\phi:\phi^2::\phi^2:\phi^3\ldots\text{etc.}$$

Progression by the $\sqrt{2}$ Division
$$1:\sqrt{2}::\sqrt{2}:2\ldots\text{etc.}$$

To analyse these two progressions we should recall some basic ideas from the grammar of our geometric language. A square figure, such as ϕ^2, represents the first plane of manifestation, that of *ideation* or *image* where a notion first becomes comprehensible. A cubic figure, such as ϕ^3, represents this same notion, idea or image in its manifest, physical, volumetric form. The inverses of these symbols $(1/\phi^2, 1/\phi^3)$ are the same principles contained within Unity; that is, they are fractions or internal parts of One, representing the preconceptual stages of these levels of manifestation. Let us remember also that One is the symbol of God. The Golden Division is the only continuous proportion that yields a progression in which the terms representing the external universe (ϕ^2, ϕ^3) are an exact, continuous proportional reflection of the internal progression $(1/\phi^2, 1/\phi^3)$ – the creative dream of God. The $\sqrt{2}$ progression on the contrary is strictly a procreative power, functioning generatively only on the external plane.

Let us again contrast the qualities of these two geometric progressions ϕ and $\sqrt{2}$ as models for evolution – progression being an apt analogy for the evolutionary process – looking this time at the phase of evolution which moves from a metaphysical, proportional principle to the physical world. The Golden Progression shows the possibility, *not* of a quantitative, statistical evolution (as with the $\sqrt{2}$ model to which Darwinian adaptation conforms), but, instead, of an evolution guided from within, an exaltation of the initial qualities of Divine ideation passing directly from the abstract into the concrete or visible; where the manifest world is an image of the Divine, a replication or son of God (Unity). The Golden Proportion represents indisputable proportional evidence of the possibility of a conscious evolution as well as of an evolution of consciousness.

St John wrote of the creative moment or original scission, 'In the beginning was the Word' (or in Greek, *logos*, meaning a three-term proportion), '. . . and the Word was with God' (the phrase 'with God' can also be read 'in God') '. . . and the Word was God'. Looking closely at these three phrases one can see that they intuitively describe the geometric implications of the Golden Proportion:

> In the beginning was the Word,
> And the Word was with[in] God,
> And the Word was God.

Workbook 5
The Golden Proportion

We begin our search for a geometric division which requires only two terms by using two geometric ideas which are already familiar: the right triangle inscribed in a semicircle (Theorem of Thales) and the $\sqrt{2}$ (Workbook 1) which in this case will be the radius of this semicircle. As shown on p. 45, we can use $\sqrt{2}$ as radius to give a division of line segments, a, b, c into a three-term geometric proportion.

Drawing 5.1a. Taking square $ABCD$, project the inner surface divisions by circular arcs onto the square's linear base. From this base line we will derive proportional relationships. With C as centre and radius CA, project base line EG. Project line CD in a similar manner, giving line DF. Using the geometric theorem that the angle inscribed in a semicircle (diameter EG) is a right angle, we join AE and AG and find three similar triangles:

$$\triangle EDA \approx \triangle EAG$$
$$\triangle EAG \approx \triangle ADG$$
$$\triangle ADG \approx \triangle EDA$$

Therefore, $a:b::b:c$,

and if $\dfrac{a}{b} = \dfrac{b}{c}$ then $b^2 = ac$.

In this case, $c = 2b+a$, and $a:b::b:2b+a$.

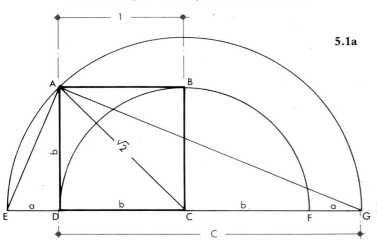

5.1a

The values shown are:
ED = $a = \sqrt{2}-1$
AB = $b = 1$
CA = $\sqrt{2}$
DG = $c = 2-\sqrt{2}$

Drawing 5.1b. We can see that the division by the diagonal in Drawing 5.1a gives a value for b which is twice that of the desired relationship: we have

$$\frac{a}{b} : \frac{b}{2b+a} \text{ as compared with } \frac{a}{b} : \frac{b}{b+a}$$

The next logical step would be to try instead the semi-diagonal as the radius of the circumscribing semicircle. This is constructed as follows:

5.1b

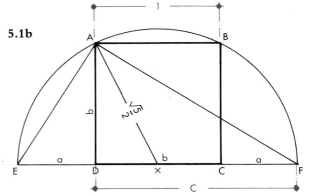

Rotate the semi-diagonal AX of square $ABCD$ to mark E and F on the extended base line. By Thales,

$$a:b::b:c.$$
$$c = a+b$$
hence, $a:b::b:a+b$

We then have the values:

side of the square $AB = b = 1$

$$XA = \frac{\sqrt{5}}{2}$$

$$ED = a = \frac{\sqrt{5}}{2} - \frac{1}{2}$$

$$DF = c = \frac{\sqrt{5}}{2} + \frac{1}{2} = \frac{\sqrt{5}+1}{2}$$

Looking at these values in a purely algebraic form,

$$\triangle DAF \approx \triangle EAD$$

Therefore $\dfrac{a}{b} = \dfrac{b}{a+b}$

and
$$b^2 = a(a+b)$$
$$b^2 = a^2 + ab$$

In this form it is evident that we have the only possible division of a unit or whole into a three-term geometric proportion which uses only two terms, an extreme term $= a$, and a mean term $= b$. This proportion was called 'the division into extreme and mean terms' and is the one named by the Greeks ϕ (phi).

Expressing this proportion as a division of 1 or unity, let $b = 1$.

Then, $\qquad b^2 = a^2 + ab$
is equal to $1^2 = a^2 + 1a$
$\qquad\qquad 1 = a^2 + a$

In substituting $b = 1$, we have $a^2 + a = 1$. This means that both a^2 and a are fractions of 1 and must therefore be written in their inverse form:

$$1 = \frac{1}{a^2} + \frac{1}{a}.$$

Drawing 5.1c. As our equation demonstrates, $a^2 + a$ fulfils the definition of the extreme and mean division of Unity. We can therefore substitute the Greek symbol ϕ for this division:

$$1 = \frac{1}{\phi^2} + \frac{1}{\phi}$$

Let us now observe this same idea in the form of tangible geometric areas. (Graph paper is useful here.) If $b = 1$ then the original square equals Unity. With D as centre, swing arc EG. With C as centre, swing arc FH.

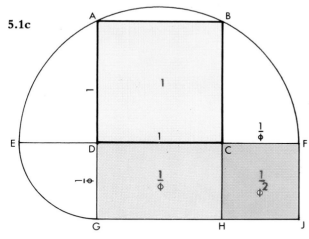

5.1c

Project GJ parallel to DC, defining the rectangle $DCHG$ and square $CFJH$.

$$\text{Area } DCHG = 1 \times \frac{1}{\phi} = \frac{1}{\phi}$$

$$\text{Area } CFJH = \frac{1}{\phi} \times \frac{1}{\phi} = \frac{1}{\phi^2}$$

– thus geometrically validating the only division of Unity into extreme and mean terms in geometric areas: $DFJG = ABCD = 1$.

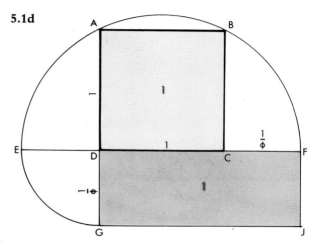

5.1d

Drawing 5.1d. Adding rectangles $DCHG$ and $CFJH$ we form the composite rectangle $DFJG$ with sides $1/\phi$ and $1 + 1/\phi$, and area 1. Therefore,

$$\frac{1}{\phi}\left(1 + \frac{1}{\phi}\right) = \frac{1}{\phi} + \frac{1}{\phi^2} = 1$$

$$\text{Area } DFJG = DF \times DG; \; DF = \frac{\text{area } DFJG}{DG}$$

$$DF = \frac{1}{1/\phi} = \phi$$

but $DF = 1 + \dfrac{1}{\phi}$

so $\qquad \phi = 1 + \dfrac{1}{\phi}$

Drawing 5.1e. Since $\phi = 1/\phi + 1$, side AG of rectangle $ABHG = 1 + 1/\phi = \phi$. Area $ABHG = 1 \times \phi = \phi$.

$ABHG$ is a Golden Rectangle.

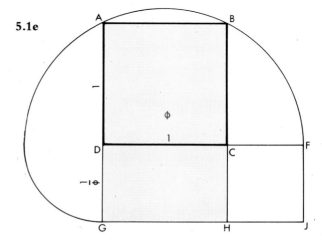

5.1e

Drawing 5.1f.

$$\text{area } BKFC = 1 \times \frac{1}{\phi} = \frac{1}{\phi}$$

$$\text{area } AKJG = \phi \times \phi = \phi^2$$

$$\text{But area } AKJG = ABCD + BKFC + CFJH + DCHG$$
$$= (1 + 1/\phi) + (1/\phi^2 + 1/\phi)$$
$$= (\phi) + (1) \text{ (by substitution)}$$

$$\text{area } AKJG = \phi^2 = \phi + 1$$
$$\phi^2 = \phi + 1$$

5.1f

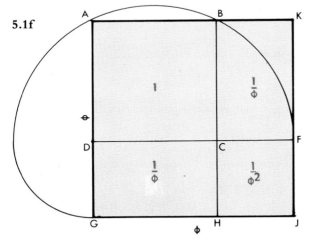

This demonstration was inspired by a similar one in *Philosophical Geometry* by André vandenBroeck.

Drawing 5.2. Geometrically, the Golden Proportion ϕ is inseparably related to the $\sqrt{5}$ function and the pentagon which we discussed in Workbook 3. It will be helpful to follow the geometry which emphasizes this relationship. Here is the method of generating the Golden Proportion from the $\sqrt{5}$ and the 1:2 rectangle.

Draw a double square and extend the dividing line *EF*. With *G* as centre and a semi-diagonal *GA* as radius, swing an arc intersecting *EF* at *H*.

$$GE = \frac{1}{2}$$

$$GH = GA = \frac{\sqrt{5}}{2}$$

$$FH = \frac{1}{2} + \frac{\sqrt{5}}{2} = \frac{1 + \sqrt{5}}{2} = \phi \text{ or } 1\cdot6180339 \ldots \text{ the}$$
$$\text{Golden Proportion.}$$

Thus the Golden Rectangle *JBFH* arises out of the double square through its $\sqrt{5}$ rectangle.

Drawings 5.3a and 5.3b. The relationship of ϕ to $\sqrt{5}$ and the pentagon. From square *ABFE* construct $HK = \sqrt{5}$. With *E* and *F* as centres and radius *FN*, swing arcs *HN* and *KN*. With *E* and *F* as centres and radius *FB*, swing arcs to intersect arcs *HN* and *KN* at *O* and *P* respectively.

$$HK = \sqrt{5}$$

$$HE + FH = \sqrt{5} - 1$$

$$HE = FK = \frac{\sqrt{5} - 1}{2}$$

$$EN = EK = \frac{\sqrt{5} - 1}{2} + 1$$

$$EN = \frac{\sqrt{5} + 1}{2}$$

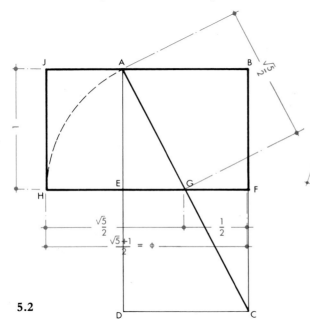

5.2

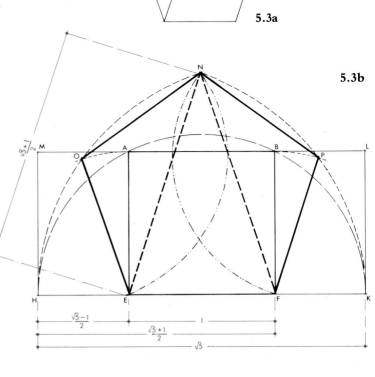

5.3a

5.3b

With the compass it can be seen that the points O, N, and P plus the two base points of the square, E and F, constitute five equidistant points. Connect F, E, O, N, P to form a pentagon.

This construction reveals an important pentagonal relationship: the side of a pentagon is in relation to its diagonal as $1 : (\sqrt{5}+1)/2$ or $1 : \phi$, the Golden Section.

Drawings 5.4a and 5.4b. These two drawings are not essential for one's understanding of ϕ, but the more enthusiastic readers will find them useful.

Drawing 5.4a. Draw a circle and cross-axes. With compass opening unchanged and with S as centre, swing an arc to intersect the circumference at 1 and 2. Join these points to determine the midpoint of the circle's radius at 3. From point 3 continue the construction as described in Workbook 3, Drawing 3.3. When the radius equals unity, the side of the inscribed pentagon, by the Pythagorean theorem, equals $\sqrt{(1 + 1/\phi^2)} = 1\cdot17557$.

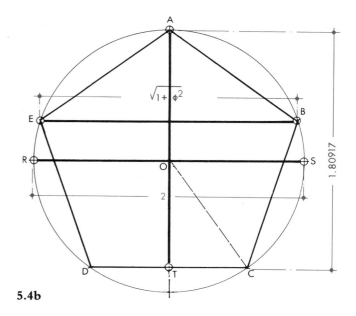

5.4b

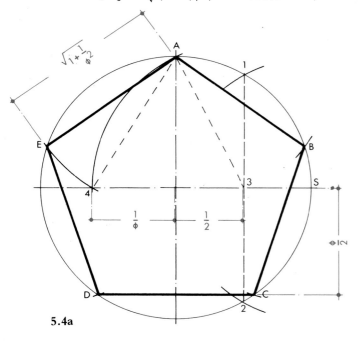

5.4a

Drawing 5.4b. Draw diagonal EB and height AT. To determine numerically the height of the pentagon we have right triangle OTC with base $TC = 1/2 \times 1\cdot17557$, that is half the side of the pentagon $= 0\cdot587785$, and the hypotenuse of triangle $OTC = OC = 1$, the radius of the circle. By Pythagoras,

$$OT^2 = 1 - 0\cdot34549 = 0\cdot65451$$

$$OT = \sqrt{0\cdot65451} = 0\cdot80901 = \frac{\phi}{2}$$

Therefore the height of the pentagon $AT = 1\cdot809$.

We have proven in demonstration 5.3a that the relationship between the side of a pentagon and its diagonal is $1 : \phi$. In the case in which the radius is 1 and the side is $1\cdot17557$ (Drawing 5.4a), the diagonal $= 1\cdot17557\phi = 1\cdot90211$.

With a radius $= 1$ or a diameter $= 2$, the diagonal $EB = \sqrt{(1 + \phi^2)} = 1\cdot90211$ and the height $AT = 1\cdot809$.

The diagonal of the pentagon is the geometric mean between the diameter of the circumscribing circle and the height of the pentagon.

$$\frac{\text{Diameter of the circle } RS}{\text{Diagonal of pentagon } EB}$$

$$= \frac{\text{Diagonal of the pentagon } EB}{\text{Height of the pentagon } AT}$$

$$\frac{2}{1\cdot90211} = \frac{1\cdot90211}{1\cdot809} = 1\cdot05147$$

The ratio 18/19 is of interest since it is one of the ratios used to define the semitone in music and is also the ratio that determines the lunar and solar year in the eclipse cycle. The ancient Egyptians based their canon for the height of man on this ratio, counting 18 units to the brow and 19 to the top of the head.

Drawing 5.5. When the *side of the pentagon* is Unity,

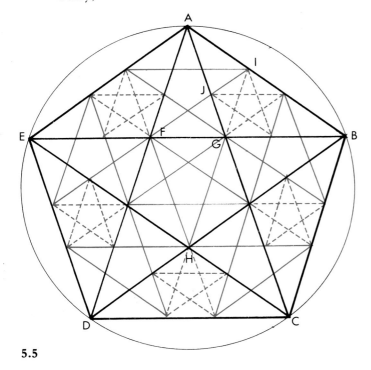

5.5

$AB = 1$

$EG = FB = 1$

$EB = \phi(1{\cdot}618)$

$GB = \phi - 1 = \dfrac{1}{\phi} \ (0{\cdot}618)$

$GI = FG = 1 - \dfrac{1}{\phi}$

$FG = \dfrac{1}{\phi^2} \ (0{\cdot}382)$

But $\dfrac{JG}{FG} = \dfrac{GB}{AB}$

$JG \div \dfrac{1}{\phi^2} = \dfrac{1}{\phi}$

thus $JG \times \phi^2 = \dfrac{1}{\phi}$

$JG = \dfrac{1}{\phi^3} \ (0{\cdot}236)$

An excellent exercise is to compute these same line segments, but starting with side $AB = 1{\cdot}17557$.

In this Workbook I have tried to encourage the reader to experience the web of modulated relationships which surround the Golden Division, ϕ. Along with the geometric demonstrations I have given the modern algebraic and decimal forms. Our argument is not to displace our modern techniques with the ancient geometric mode, but to reground our numerical language in the visual and spatial world from which it arose.

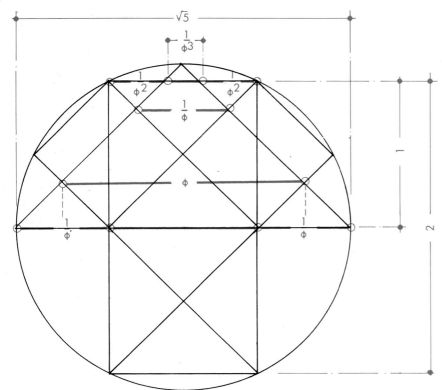

The $\sqrt{5}$ and its double square rectangle inevitably generate or disclose an array of proportions associated with the Golden Mean or Golden Proportion.

Johannes Kepler, formulator of the laws of planetary motion, is quoted as saying, 'Geometry has two great treasures: one is the theorem of Pythagoras, the other the division of a line into mean and extreme ratios, that is ϕ, the Golden Mean. The first way may be compared to a measure of gold, the second to a precious jewel.'

There are grand philosophical, natural and aesthetic considerations which have surrounded this proportion ever since humanity first began to reflect upon the geometric forms of its world. Its presence can be found in the sacred art of Egypt, India, China, Islam and other traditional civilizations. It dominates Greek art and architecture; it persists concealed in the monuments of the Gothic Middle Ages and re-emerges openly to be celebrated in the Renaissance. Although it pervades many aspects of nature from which the artists drew their inspiration, it would be wrong to say that one can uncover the Golden Mean throughout all of Nature. But it can be said that wherever there is an intensification of function or a particular beauty and harmony of form, there the Golden Mean will be found. It is a reminder of the relatedness of the created world to the perfection of its source and of its potential future evolution.

The Golden Divisions contained in the pentagram are shown to determine the proportions of this ancient mask of Hermes.

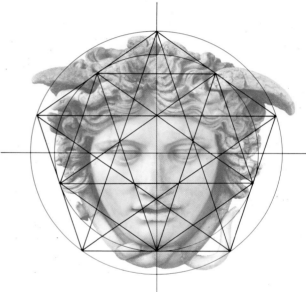

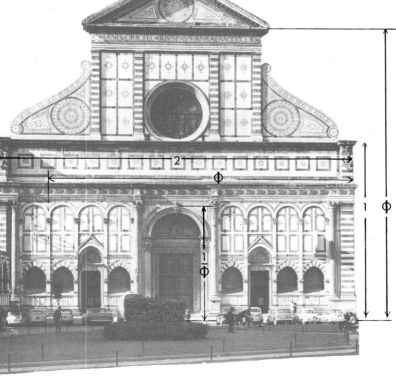

Because of the distortion of perspective inevitable in a photograph, we can only roughly indicate a few of the basic ϕ proportions. But this entire edifice is based on ϕ and $\sqrt{2}$ relationships.

THE TOMB OF PETOSIRIS

This Egyptian tomb of the Ptolemaic period was discovered in 1919 and excavated under the direction of Gustave Lefebvre who published his findings in 1924. It is near the city of Hermopolis, the city of Thoth, in a necropolis not far from the underground cemetery of sacred ibises, the animal sacred to Thoth. It was built around 300 BC for Petosiris and his family, including his father, step-father, brothers, wife and children. All the men of the family bear the titles 'Great of the Five', and 'Master of the Seat', which are the titles of the High Priest of Thoth of Hermopolis.

The name Petosiris means 'gift of Osiris'. The builder of this tomb was evidently an exceptional man, for half a century after his death he was ranked with Imhotep and Amenhotep Son of Hapu as a semi-divine sage, and his tomb was a place of pilgrimage.

Painted bas relief from the east wall of the tomb's chapel. The priest pours a libation over the mummy of the deceased.

Geometrical analysis with the base of the triangle *EC* as unity. The construction of the square *KLCE* and the semi-diagonal *PK* verifies that both *AC* and *CG* = ϕ.

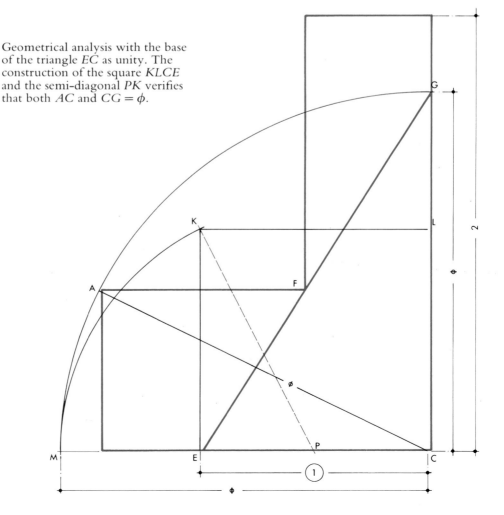

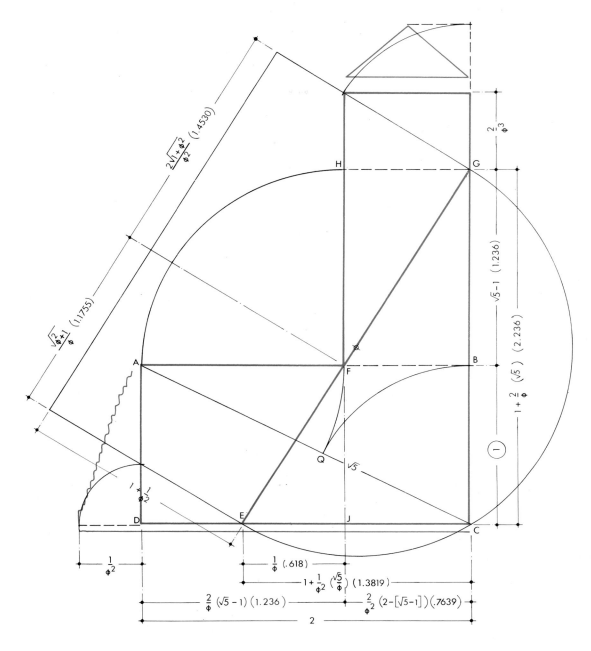

$\frac{2\sqrt{1+\phi^2}}{\phi^2}$ (1.4530)

$\frac{\sqrt{2+1}}{\phi}$ (1.755)

$\frac{2}{\phi^3}$

$\sqrt{5}-1$ (1.236)

$\frac{2}{\phi}(\sqrt{5})$ (2.236)

$1+\frac{2}{\phi}(\sqrt{5})$

(1)

$1+\frac{1}{\phi^2}$

$\frac{1}{\phi^2}$

$\frac{1}{\phi}$ (.618)

$1+\frac{1}{\phi^2}(\frac{\sqrt{5}}{\phi})$ (1.3819)

$\frac{2}{\phi}(\sqrt{5}-1)$ (1.236)

$\frac{2}{\phi^2}(2-[\sqrt{5}-1])$ (.7639)

2

Geometrical analysis with the height of the horizontal rectangle BC equal to unity. An arc is swung from C with radius CB to Q, and likewise with centre A and radius AQ, determining FJ to be a Golden Mean division of the horizontal rectangle.

By the Pythagorean theorem,

$$EF^2 = EJ^2 + JF^2$$
$$EF = \sqrt{[(1/\phi)^2 + 1^2]}$$
$$= \sqrt{(1/\phi^2 + 1)}$$
$$= [(\phi^2 + 1)/\phi] = 1$$

(side of the inscribed pentagon to circle with radius 1)

$$FG^2 = JC^2 + BG^2$$
$$= (\frac{2}{\phi^2})^2 + [(1+\frac{2}{\phi}) - 1]^2$$
$$= \frac{2\sqrt{(1+\phi^2)}}{\phi^2}$$

(side of the circumscribed pentagon to circle with radius 1)

Also, $1 \cdot 17557 + 1 \cdot 453085 = 2 \cdot 628655$, which is the side of the rectangle $EE'GG'$ whose width is

$1 \cdot 3819$, and $2 \cdot 6286/1 \cdot 3819 = 1 \cdot 9022 = \sqrt{(1+\phi^2)}$, or the ratio of the side (1) of a pentagon to its diagonal. In comparing one system of proportions to the other, we find

$$\frac{\text{Unity } (EC)}{\text{Unity } (CB)} = \frac{AC}{ac} = \frac{\sqrt{5}}{\phi} = \frac{2 \cdot 236}{1 \cdot 618}$$
$$= 1 \cdot 38196 = 1 + \frac{1}{\phi^2}$$

This analysis shows that the Master Petosiris had a complete and extremely sophisticated knowledge of the Golden Proportion revealed very simply in a play of geometric relationships resulting from two overlapping rectangles. The Golden Mean proportion philosophically represents the seat or basis of the created worlds, hence, perhaps, the title, 'Master of the Seat'. The burial practices in the Pharaonic tradition were undertaken not merely to provide a receptacle for the physical body of the deceased, but also to make a place to retain the metaphysical knowledge which the person had mastered in his lifetime. The proportions of the seat of Petosiris as shown in his tomb reflect this intention.

It is important to mention, first of all, that ϕ represents a coinciding of the processes of addition and multiplication. Addition is the most common process of growth, whether it be of cells in our body, of wealth, of knowledge, or of experience; it is a deliberate, logically expanding development. Multiplication is really a special form of addition, an accelerated form: 4×4 is really $4+4+4+4$. But in this acceleration there is the intervention of an extraordinary moment of transformation: what was a linear accumulation suddenly becomes a square, a surface, a plane. There has been a leap of growth. In the plant the simple additive growth occurring in the stem suddenly explodes into a fruit or flower, or a seed gradually swells from absorbing moisture and germinates. In studies, one's additive accumulation of skills or data suddenly blossoms into a genuine understanding. The clearest observation of this moment occurs in the process of growing a crystal. One gradually adds a mineral salt to a small dish of water over a period of days. The water dissolves the salt, but at the same time the air is slowly evaporating or subtracting the water. When the point of saturation is reached, and this is amazing to watch under a microscope, the so-called 'mother tincture' suddenly congeals into a geometrized expression of the salt as a crystal. When such a moment occurs in the context of spiritual development it is called redemption or enlightenment.

There are three significant circumstances in which the ancient researchers of this principle found this simultaneous coincidence of additive and multiplicative processes. Each of these gives the sense of a combination of material and supramaterial growth. They are the square (which we saw in Workbook 1), musical harmony (Workbook 8) and the proportion ϕ.

The cube of phi, ϕ^3, is a volume arrived at by simultaneously adding and multiplying.

$$\frac{1}{\phi} + 1 = \phi = 1 \times \phi$$
$$1 + \phi = \phi \times \phi = \phi + 1$$
$$\phi + \phi^2 = \phi^3 = \phi \times \phi \times \phi = \phi \times \phi^2$$

The volumetric expression of ϕ, ϕ^3 becomes the new unity, for here the abstract principle of ϕ achieves expression as a unity on the physical level of volume, the cube. In an ancient Egyptian inscription Thoth says,

I am One which transforms into Two	polarity
I am Two which transforms into Four	surface, $2^2 = 4$
I am Four which transforms into Eight	volume, $2^3 = 8$
After all of this, I am One.	

The progression then occurs as though we were to continue to consider the One as without definition, up until the moment it becomes a tangible, manifest unit, the cube; as we've just seen, $\phi^3 = 1$. And if the transformative power of redemption is fixed to the material cross, the cross of addition $+$, then the moment of resurrection comes when this principle allows the cross to fall $+ \times$, and an exponential growth occurs, an incomprehensible, non-sequential leap to another level of being.

We will see in the next chapter the forms of exponential growth exemplified in the logarithmic spirals based on the roots of 2, 3, and 5. The Golden Mean spiral, in which the geometric increase of the radial arms is equal to ϕ, is found in nature in the beautiful conch shell *Nautilus pompilius* which the dancing Shiva of the Hindu myth holds in one of his hands as one of the instruments through which he initiates creation. To Pythagorean eyes, however, this form embodies the dynamics of the rhythmic generation of the cosmos, and through its harmonic principle, represents universal love. The logarithmic spiral is found to be superimposable on the foetus of man and animals, and is present in the growth patterns of many plants. The distribution of seeds in the sunflower, for example, is governed by the Golden

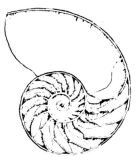

Nautilus pompilius.

Mean logarithmic spiral. The sunflower, furthermore, has 55 clockwise spirals overlaid onto either 34 or 89 counterclockwise spirals. We recognize these numbers as part of the Fibonacci Series, which is generated by ϕ.

Series A   1, 1, 2,   3,   5,   8, 13, 21,   34,   55,   89, 144,   233,   377,   610

Series B   1, 3, 4,   7, 11, 18, 29, 47,   76, 123, 199, 322,   521,   843, 1364

Series C   1, 5, 6, 11, 17, 28, 45, 73, 118, 191, 309, 500,   809, 1309, 2118

Series C^1   Series C × 2 ——————— 236, 382, 618,   1, 1618, 2618, 4236

Series D ϕ^{-5}, ϕ^{-4}, ϕ^{-3}, ϕ^{-2}, ϕ^{-1}, 1, ϕ^1, ϕ^2, ϕ^3, ϕ^4, ϕ^5
0·090, 0·1458, 0·236, 0·3819, 0·618, 1, 1·618, 2·618, 4·236, 6·854, 11·090

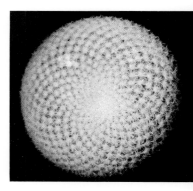

Seed distribution in a cactus plant which, as in the case of the sunflower, conforms exactly to the Golden Spiral.

The series of numbers called 'Fibonacci' is a special additive progression in which the two initial terms are added together to form the third term (*Series A*). For example:

first term = 1
second term = 1
third term = 1 + 1 = 2
fourth term = 1 + 2 = 3
fifth term = 2 + 3 = 5 . . . etc.

The Fibonacci Series is such that any two successive terms tend to be approximately in relation to one another as $1 : \phi$, and any three successive terms are as $1 : \phi : \phi^2$ . . . etc. Let us take, for example, the tenth and eleventh terms from Series A:

$$\frac{89}{55} = 1 \cdot 61818 = \text{approximately } \phi$$

$$\frac{144}{55} = 2 \cdot 61818 = \text{approximately } \phi^2$$

Although the Fibonacci, the most common additive series, begins 1, 1, 2 (note the similarity in this respect with Theon's series, which we saw in Chapter IV), it is possible to begin an additive series with any two ascending numbers, for example, *Series B*, 1, 3, 4, 7, etc. In every series of this type the successive ratios will tend toward ϕ, and it is interesting to note that the relationship between the corresponding terms in two series A and B tends toward $\sqrt{5}$. For example, with the twelfth term from Series A and from Series B,

$$\frac{521}{233} = 2 \cdot 23605 = \text{approximately } \sqrt{5}$$

In *Series C* the additive progression beginning 1, 5, 6, 11 has the mystifying characteristic that the whole numbers themselves tend to be exactly half of the decimal expression of the Golden Ratio. For example, the twelfth term of Series C = 309, and 309 × 2 = 618, while $1/\phi = 0 \cdot 6180337$. . . .

Series D shows how the Golden Series is the model progression for the logarithmic principle in which there is a relationship between an additive series ('exponents') and a multiplication series ('terms') such that by simply adding the exponents one can determine the corresponding multiplication of terms. For example,

$$2^2 \times 2^3 = 2^5$$
$$\phi^2 \times \phi^3 = \phi^5$$

The multiplication of the numbers, or in this case decimal terms, is equal to the addition of the exponents.

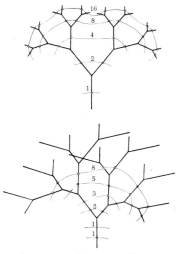

The two major branching patterns, one demonstrating the geometric progression by 2 ($\sqrt{2}$), and the other the Fibonacci Series (ϕ).

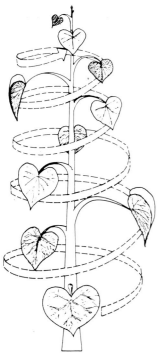

The distribution of leaves around a central stem is governed by the Fibonacci Series: 3 leaves in 5 turns, 5 leaves in 8 turns.

Named after the thirteenth-century Italian mathematician who drew attention to it, the Fibonacci Series appears in many places in natural phenomena, and a number of studies document its ubiquity. It governs, for example, the laws involved with the multiple reflections of light through mirrors, as well as the rhythmic laws of gains and losses in the radiation of energy. The Fibonacci Series perfectly delineates the breeding pattern of rabbits, a symbol for fecundity, and the ratio of males to females in honey bee hives. Philiotaxis is the botanical term describing the arrangement of leaves on the stem of a plant. If a helix is drawn passing through each leaf base until it reaches the first base which is vertically above the starting point, and P is the number of turns of the helix and Q is the number of leaf bases passed, then P/Q is a fraction which is characteristic of the plant's leaf-distribution pattern. Both the numerator and denominator of this fraction tend to be members of the Fibonacci Series A. Naturally a botanist's interest in this distribution is not primarily mathematical. His attention is directed to the fact that all the members of this series of fractions lie between 1/2 and 1/3, creating the situation in which successive leaves are separated from one another by at least one-third of the stem circumference, thus insuring a maximum of light and air for the leaf which is below.

Branching is another major functional pattern of natural growth which is controlled by the Fibonacci or ϕ series. And because of its appearance in the pentagon, the Golden Section can be found in all flowers having five petals or any multiple of five, and the daisy family will always have a number of petals from the Fibonacci Series. The rose family is one of those based on five, as are all the flowers of the edible fruit-bearing plants. Thus five signals to man his proper foods. Five is dominant in the substructure of living forms, while 6 and 8 are most characteristic of the geometry of mineral, inanimate structures. The plants displaying a sixfold structure, such as the tulip, the lily and the poppy, are very often poisonous or only medicinal for man. Traditional medicine considered seven-petalled plants to be poisonous. Among these are the tomato and other plants of the belladonna or nightshade family. The very exotic flowers, on the other hand, the flowers of love such as the orchid, the azalea and the passion flower, are all governed by pentagonal symmetry. The pentagon as the symbol of life, particularly of human life, was the basis of many Gothic rose-window mandalas.

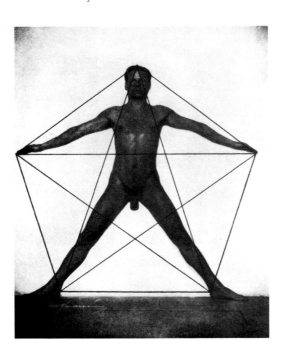

Five as the flowering or quintessence of life.

Man as the pentagon.

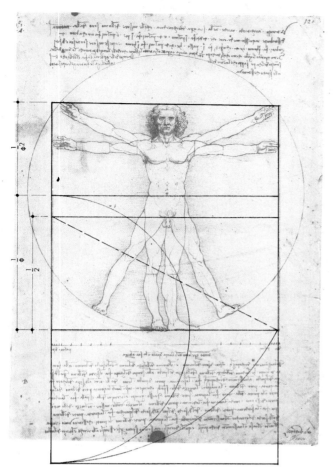

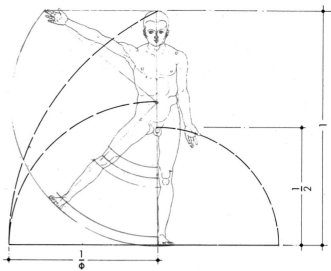

The canonical figures of both Leonardo da Vinci and Albrecht Dürer conform to the ancient biometric symbol of the body divided in half by the sex organ and by ϕ at the navel.

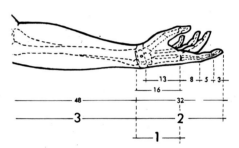

The appearance of the Fibonacci Series in the relationships between the bone-lengths of the human finger, hand and arm is another instance of the numerous ϕ relationships which occur in the human body.

It is however in the human body that we can uncover the metaphysical meaning of ϕ as implied in the dictum of Heraclitus, 'Man is the measure of all things'. According to several traditions which provide us with a human canon, that is a delineation of the average and ideal body proportions, the navel divides the body according to the Golden Section. Taking the full height as 1, the body from the feet to the navel, in the Egyptian, Greek and Japanese canons, is equal to $1/\phi$, with the portion from the navel to the top of the head equal to $1/\phi^2$. The body is divided exactly in half by the sex organs. This denotes the relationship of sexuality with the dualizing function, the division into two. At birth, however, it is the *navel* that divides the child exactly in half, and in the course of maturation the navel moves to the point of the phi division. Thus the position of the navel through human growth is related to the idea of a movement from the dualized, sexualized stance in nature to that of a proportional relation to Unity through the asymmetrical, dynamic power of ϕ.

The study of human biometrics reveals a nuance to this proportioning. In the female the navel is normally a little above the exact cut of the Golden Section while in males it is a little below. Furthermore, during the growth process of both males and females the placement of the navel falls sometimes above, sometimes below the ϕ division of the body. This shifting begins at puberty and recurs between the ages of 17 and 30. Such an oscillation above and below an irrational point of formative perfection is a principle we also find as the basis of ancient mathematics: as in the Diaphantine method where whole number ratios approach in progression the sacred or incommensurable root functions.

The Osirion is a large, underground Egyptian temple which is an architectural allegory depicting the process of transformation through death and rebirth as rendered in the myth of Osiris. The symbolism of Osiris is concerned with cyclic rebirth and transformation on both individual and universal levels, and the Osirion was designed to represent the tomb itself of Osiris. This temple may or may not have functioned as an initiatic temple, but its architecture is symbolic, in every detail, of the mechanics of reincarnation, whether that be in reference to physical death and rebirth, or the death of one phase of consciousness in the seeker and birth into a new, or the death and dissolution of the universe and its return.

The Osirion was discovered at Abydos in 1901 by Flinders Petrie, and excavations were completed in 1927. It is considered to be the cenotaph (empty tomb) of Seti I who ruled Egypt from 1312 to 1298 BC. The entire temple was roofed over and then a huge mound of earth was placed on top so that it resembled an underground tomb. Around the buried temple huge pits were dug and the tree sacred to Osiris was planted. The image from a sarcophagus here shows the symbol of the tomb of Osiris with the trees of rebirth shooting from it.

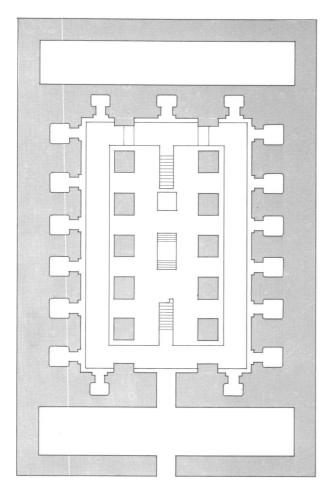

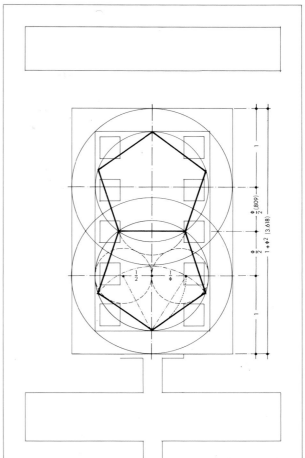

The plan of the Osirion shows a very curious central area with ten large, square columns (shaded) which support the roof. This platform, with steps leading up to it on both ends, is actually an island, for it is surrounded by an aisle dug out to the precise level to enable it to fill with ground water. The island with steps on both sides exactly resembles the Egyptian symbol for the primordial hill or mound which, in the myth, represents the first place of creation which arises out of the primeval waters, the unmanifest, undifferentiated *Nun*. Osiris also represents the principle of the seed buried in the soil which germinates upon absorbing moisture from the ground.

There are three burial locations in this symbolic tomb, two depressions on the central platform (one presumably for the coffin and the other for the canopic vases) and a long, sealed burial chamber, itself shaped like a sarcophagus, at the far end. The latter contains astronomical wall and ceiling carvings, bringing celestial influences into the tomb. Around the outside of the central hall are seventeen small chambers. It is speculated that perhaps these rooms were for neophytes who went through an initiatic rite of descending into the watery depths and re-emerging on the central island, symbolizing the mystery of rebirth on both universal,

cosmic and individual levels (provided of course that there was air to breathe in the tomb). In any case, and more importantly, the geometry of the temple supports this theme by conforming to the proportions of the Golden Section and $\sqrt{5}$, the symbol of rebirth and regeneration, as well as to the $\sqrt{2}$, the symbol of the procreative, self-generative power of life. The emphasis on the theme of the pentagon aptly symbolizes the belief that the king, after death, became a star (the star was always represented in Egypt as having five points). (The geometric analyses of both the Osirion and the Tomb of Petosiris have been graciously contributed by Lucie Lamy.)

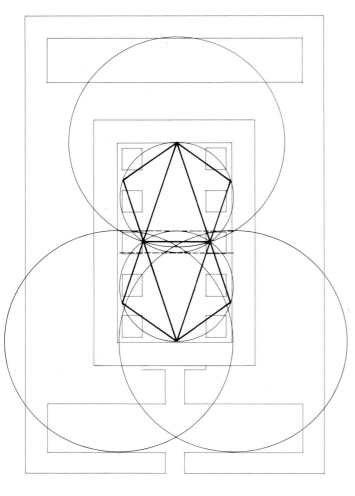

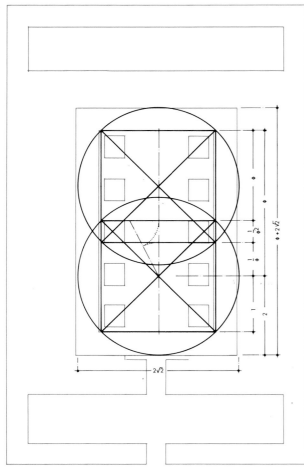

$\phi + 2\sqrt{2}$

$\frac{1}{\phi^2}$

$\frac{1}{\phi}$

$2\sqrt{2}$

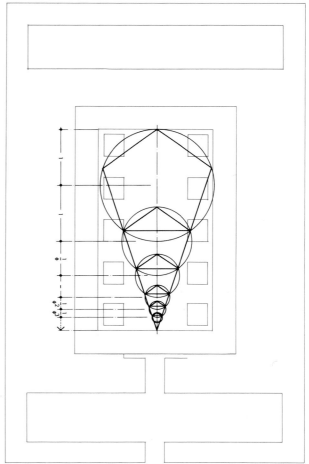

We can summarize some of the ideas evoked by this most fundamental proportional relationship as follows. As the ancients say 'The universe is God regarding himself'. Creation cannot exist without perception, and perception is relationship: 'To be is to relate.' The archetypal patterns of relationship can be contemplated through the laws of proportion contained in pure number and geometric form. The Golden Proportion is the transcendent 'idea-form' which must exist *a priori* and eternally before all the progressions which evolve in time and space.

This drawing shows the geometric allegory for the concept of the Holy Trinity, the Three who are One. $1 =$ God the Father; $1/\phi =$ The Holy Spirit (the binding function, or *prana*); $1/\phi^2 =$ The Son (The Square or potential for manifestation, the Supreme Archetype).

These terms form a three-term proportion:

$$\frac{1/\phi^2}{1/\phi} :: \frac{1/\phi}{1}$$

Thus the extreme terms are in an identical relation to one another: Father and Son joined through the Holy Spirit. $1/\phi^2$ is the ideation of divine manifestation. $1/\phi^3$ is the embodied universal individual, the Christos. The crossing or overlapping of $1/\phi$ with $1/\phi^2$ produces $1/\phi^3$, the embodiment of Divine Man.

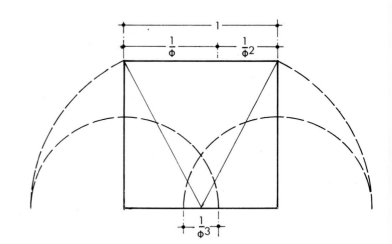

Piero della Francesca's *Baptism of Christ* follows the geometric symbolism of the Golden Proportion as the Holy Trinity. The body of Christ is exactly contained in the area $1 \times 1/\phi^3$. The Holy Spirit is contained in the overlapping of $1/\phi$ with $1/\phi^2$, and touches or binds together two areas equal to $1/\phi^2$. The height of Christ is $3 \times 1/\phi^3$.

The logic for the association of Christ (the redemptive principle) with the Golden Mean proportion is thus very directly stated here: Christ is 'the Word made flesh'. 'The Word' is the English translation of the Greek *logos*, which is defined as a continuous proportion in which the seemingly irreconcilable extreme terms are bound or interrelated by a single mean term, $a:b::b:c$. Christ as *logos* links the extreme terms of spirit and matter, universal and individual, infinite and finite.

The geometry of this painting also indicates the 'vital centres' in the body of Christ, symbolizing the 'path of resurrection' in the Tantric tradition, for the function of the vital centres in the path of spiritual unfoldment was recognized in Christian as well as in Eastern mysticism. Here the pubic arch falls at $1/\phi^2$ from the feet. The centre below the navel is indicated by the crossing of the $\sqrt{2}$ lines. The heart centre is given by the crossing of the two ϕ arcs, and the baptizing hand of John indicates the crown centre at a distance of $1/\phi^2$ from the navel centre.

The illumination depicts
creation and evolution (the
six days of Creation) through
a combination of $\sqrt{2}:1$,
which are the proportions of
the overall page, and $1:\phi$,
which is the portion con-
taining the Creation in six
stages. The Father, Son and
Holy Spirit preside over
Creation as the principle of
the Three that are One.
Always, in sacred literature,
Creation and evolution are
contemplated through the
image of the Trinity and the
two generative proportions.

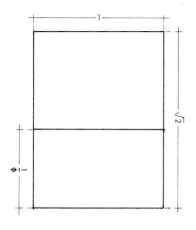

The growth of the human body describes a relation between two generative
powers, that of the $\sqrt{2}$, resulting from halving then doubling, indicated by the
location of the sex organs at the body's midpoint, denoting the quantitatively replicating *procreative* principle; and that of ϕ, indicated by the navel, denoting the
relational power which integrates the parts with each other and with the all-containing wholeness, as the umbilical point relates the child to its origin – mother universe.
In this way ϕ becomes the geometric symbol for the idea of Christ, that which binds
together the individualized awareness with the ideal totality from which it originated and toward which it will necessarily return.

> I am that which binds,
> I am the golden navel of the universe.
> He who knows this knows Upanishad.

(Upanishad means 'near approach'.)

VI Gnomonic Expansion and the Creation of Spirals

'There are certain things,' said Aristotle, 'which suffer no alteration save in magnitude when they grow . . .' He was referring here to the phenomenon the Greek mathematicians called the *gnomon* and the type of growth based upon it, known as gnomonic expansion. Hero of Alexandria defined it as follows: 'A gnomon is any figure which, when added to an original figure, leaves the resultant figure similar to the original.' The contemplation of this figure leads to an understanding of one of nature's most common forms of growth, growth by accretion or accumulative increase, in which the old form is contained within the new. This is the way the more permanent tissues of the animal body, such as bones, teeth, horns and shells, develop, in contrast to the soft tissue which is discarded and replaced.

This familiar kind of growth has often been presented architecturally as the design theme of a building. The Hindu temple is an excellent example of this. The floor was begun by placing together four bricks, each one foot square, forming thus the square of 2, then expanding this platform to the square of 3 and so forth. Each sequential expansion was considered as an expansion of the altar of sacrifice, in that the whole temple recapitulated its essence-seed, the altar, or original square. Thus the building itself expressed the meaning of 'sacrifice', which implies a reduction to that which is sacred. Both in plan and in volume the typical Hindu temple displays the kind of gnomonic growth most obviously displayed in seashells, where the residues of the previous stages of growth remain clearly indicated as part of the structure and design of the subsequent stages.

The gnomonic expansion or increase depicted in various geometric figures, and by unity dots in the form of square, rectangle, triangle.

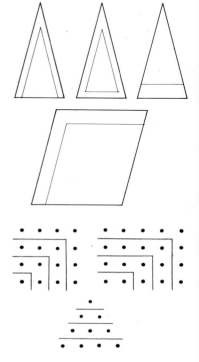

This method of figuring the gnomon shows its relationship to the Pythagorean formula $a^2 + b^2 = c^2$. Shown here is the gnomonic increase from the square surface area of 4 to the square of 5, where the gnomon of the larger square 5 is equal to 1/4 of the initial square of 4.

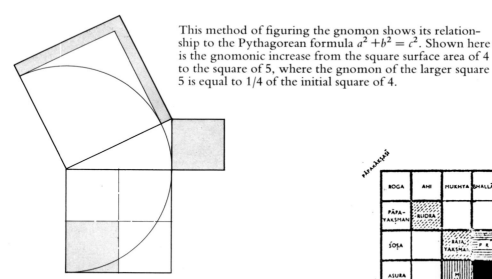

The floor plan of the typical Hindu temple is a simple, concentric gnomonic expansion of an initial square. As the mandala reflects the celestial order, each square contains the name of a deity.

The gnomon, as a succession of increments of growth, defines a passage through time. This expansion, in the Hindu temple, is an extension of the initial square which is the altar of sacrifice, the container of the symbolic cosmic fire. So time is depicted as a relentless expanding fire of life, throwing outward and consuming again the forms held as potential in the initial seed altar.

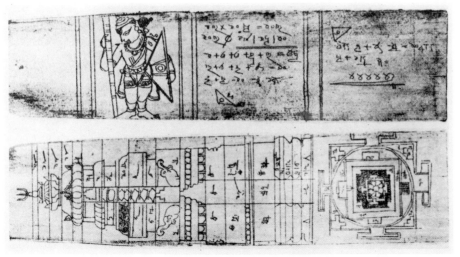

The gnomonic mandala of the floor plan is also used as the guiding element for designing the elevation of the temple.

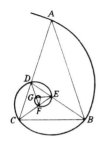

There are interesting developments of growth and number through gnomonic expansion. One mathematical characteristic is that all figures which grow by gnomonic expansion create intersections upon which spirals can be drawn. These forms, as Jill Purce has so beautifully shown in *The Mystic Spiral*, are everywhere in nature: spiralling trunks of huge eucalyptus trees, the horns of rams and reindeer, our skeletal bones, mollusc shells, particularly the *Nautilus pompilius* which follows the spiral derived from the Golden Proportion. Spirals can be found in the successive florets of the sunflower, in the outline of a cordiform leaf; in a lock of hair or a snake coil or an elephant trunk, an umbilical cord or in the cochlea of the inner ear.

All these spirals are a result of the process of gnomonic growth, of which the square and its gnomon can be considered the archetypic form.

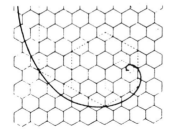

These diagrams from D'Arcy Thompson's *On Growth and Form* show that spirals can be drawn from the gnomonic growth of triangles and hexagons.

Workbook 6
Gnomonic spirals

The following demonstrations give an insight into the ancient mathematical method for generating whole number ratios which closely approximate to incommensurable functions. This method is attributed to the Greek mathematician Diaphantus, but is probably part of a much older mathematical knowledge. We can find in these demonstrations the integration of gnomonic growth, the important additive number progressions, the progression of sacred rectangles, and the number ratios which approach the sacred roots of 2, 3, and 5. All these geometrical operations become the basis for the formation of spiral curvatures which serve as a model for a vast range of universal movement, from particle to galaxy.

We begin with two additive progressions (already encountered in relation to ϕ, p. 57). We can observe how these same number series can also be conceived as a progression of expanding (whirling) rectangles in spiral formation. Our method will be to compare the relationships between the progressions emerging from the two essential creative relationships, 1:2 and 1:3. In order to do this, one series will be considered as a succession of numerators and the other as a succession of denominators. We shall begin with the formation of a spiral based on $\sqrt{5}$.

$$\frac{\text{Origin } 1:3}{\text{Origin } 1:2} \quad \frac{1 \quad 3 \quad 4 \quad 7 \quad 11 \quad 18 \quad 29 \quad 47 \quad 76 \quad 123}{1 \quad 1 \quad 2 \quad 3 \quad 5 \quad 8 \quad 13 \quad 21 \quad 34 \quad 55}$$

There are two characteristics of these progressions of fractions that should be emphasized.

First, the higher we go in the series the more closely the relationship between the numerator and the denominator approaches the incommensurable root of 5, 2·2360679 . . . .

For example, the function from our series, $29/13 = 2·230$ . . ., is an approximation a little below the $\sqrt{5}$. But the next fraction, $47/21 = 2·23809$ . . ., is an approximation this time a little above the true value of the $\sqrt{5}$. The following fraction, $76/34 = 2·235$, is again below the incommensurable root but much closer than the previous ratio; $123/53 = 2·23636$ is above yet still nearer the desired ratio. The pattern is again an oscillation above and below, ever approaching nearer the supra-rational root.

The second characteristic is that we can conceive of these successive numerical relationships as spatial forms, that is, squares and rectangles. To transform this series into a spatial configuration we simply consider 1 as the side of a square area, and add a succession of squares to our existing figure with the side of each new square equal to the preceding expansion of the initial figure:

1, 1, 2, 3, 5, 8, 13, 21, 34, 55, 89, 144, etc.

The originating 1:2 rectangle is already formed from $1+1$, so the length 2 becomes the side of a square which is added to the original rectangle giving 3. This 3 becomes the side of a new square which is added to the preceding 3:2 rectangle, giving the new relationship, 3·5.

The relationship between two successive numbers of this series tends to approach ϕ. The logarithmic function of ϕ (see p. 56) allows one to find, by taking three successive numbers of this series, the unity ϕ^2 for example, since $1+\phi = \phi^2$ exactly as the sum of 8 and $13 = 21$, and the relationship 13:8 can be considered as a value approaching that of ϕ, while 21:8 is a value of ϕ^2.

We now take our series of numerators and transform them into a spatial configuration by considering the relationship 1:3 as a rectangle, and proceed as before by the addition of a square. The first square to have a side of 3 when added to the original rectangles gives the relationship 3:4. The second square will have 4 as its side, which when added to $3 = 7$, giving the second relationship of 4:7. In continuing in this way we form the series of numbers,

1, 3, 4, 7, 11, 18, 29, 76, 123, 199, etc.

This is another sequence of numbers, different from the one attributed to Fibonacci, but again the relationship between successive terms tends toward ϕ, and each is formed by the addition of the two preceding terms.

We can now reunite these two series, each of which has ϕ for the relationship between each of its successive terms, and which between them define $\sqrt{5}$. The spiral is formed from this union.

Using this method we can develop the plan for three spirals whose curvatures express these geometric and proportional laws.

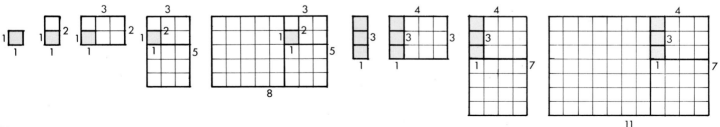

Drawing 6.1. The root 5 spiral, starting from ratios 1:2 and 1:3.

For the $\sqrt{2}$ spiral we begin again with the two creative relationships 1:2 and 1:3 to initiate progressions which will form the numerators and denominators of a series of fractions:

$$\frac{\text{origin 1:3} \quad 1 \quad 3 \quad 7 \quad 17 \quad 41 \quad 99}{\text{origin 1:2} \quad 1 \quad 2 \quad 5 \quad 12 \quad 29 \quad 70} \quad \frac{\text{diagonal numbers}}{\text{lateral numbers}}$$

Here we find two variations from the $\sqrt{5}$ formation described above. In this instance neither progression begins with the repeated number 1, and here we have, instead of the simple additive series, an addition each time of the sides of two squares.

The growth is made by the addition of two similar squares having the larger side of the preceding rectangle for side. Thus to the original 1:2 rectangle add two squares having 2 for their sides, to give a side of $1+2+2=5$; then, to the 2:5 rectangle, add two squares of side 5 which makes $2+5+5=12$, etc.

To the original 1:3 rectangle we add two squares of side 3, making $1+3+3=7$, and to this 7 we add two squares of side 7, that is $3+7+7=17$, etc. The series 1, 2, 5, 12, 29 ... etc. represents the sides of the squares in which the diagonals are respectively 1, 3, 7, 17, 41 ... etc. The ratio between these two series, moving away from unity ('the unity', as Theon put it, 'being virtually the side and the diagonal'), approaches closer and closer to $\sqrt{2}$.

6.1

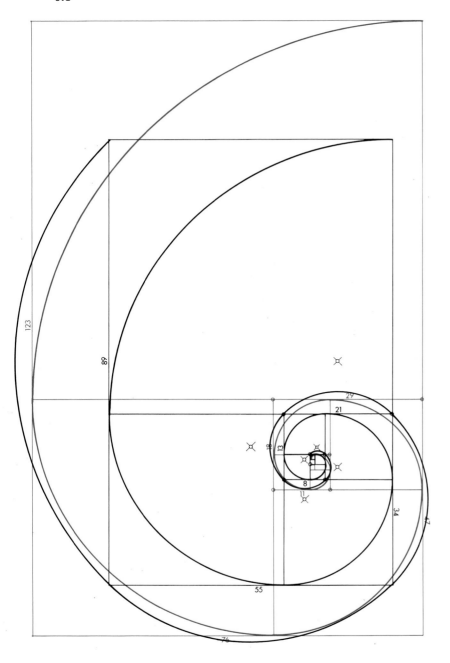

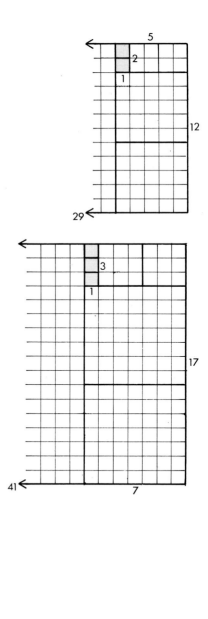

Drawing 6.2. The $\sqrt{2}$ spiral, starting from ratios 1:2 and 1:3, but with the successive addition of two squares.

With only a few modifications in the general procedure we can now construct the progression and the spiral related to the $\sqrt{3}$. The variations are that in this case the relationship 1:3 begins 1, 1, 3 . . . (rather than 1, 3 . . .) and provides the denominators instead of the numerators as it did in the other two spirals. For the $\sqrt{5}$ spiral we successively added one square, and for the $\sqrt{2}$ spiral we successively added two squares, whereas in this case we will add first two squares then one square.

$$\frac{\text{Origin } 1:2}{\text{Origin } 1:3} \quad \frac{1 \ \ 2 \ \ 5 \ \ 7 \ \ 19 \ \ 26 \ \ 71 \ \ 97}{1 \ \ 1 \ \ 3 \ \ 4 \ \ 11 \ \ 15 \ \ 41 \ \ 55} \text{ etc. or } \frac{\sqrt{3}}{1}$$

In starting from the origin 1:2 we add two squares of side 2 to total $1+2+2 = 5$, then a single square of side 5 to make $2+5 = 7$ etc., continuing to follow this alternation of the addition of two, then of one square.

The originating figure 1:3 is constructed exactly in the same way and provides the series of numbers listed above.

As in the case of the first two roots, it is the superimposition of the numerators and the denominators which provides the relationships constituting the $\sqrt{3}$. Because of the 'syncopated' addition, of first two then one square, it is impossible in this construction to draw both the internal and the external spirals. The $\sqrt{3}$, being the formative principle, acts only as the containing or external spiral.

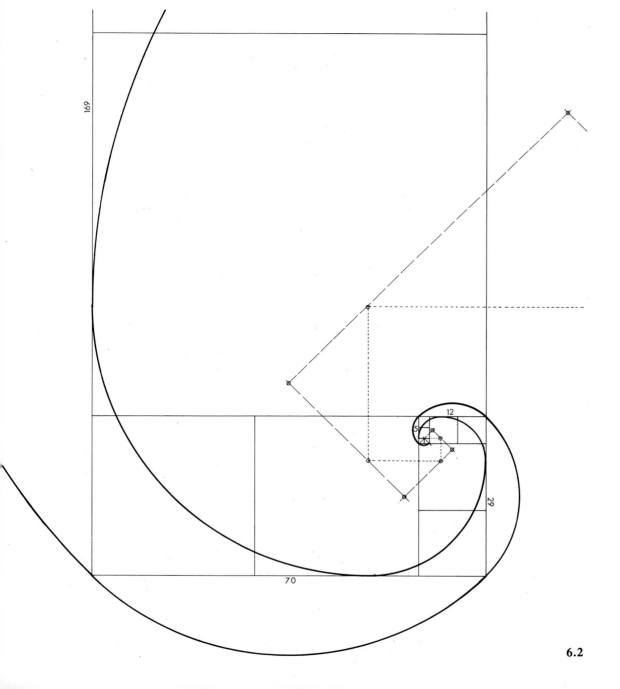

6.2

Drawing 6.3. These demonstrations of the construction of the spirals have been taken in part from *Le Temple de l'homme*, by R.A. Schwaller de Lubicz.

The deeper purpose for this development of the spiralling of numbers around the supra-rational roots comes from the fact that we have a model for the way an indefinable cause (root) can express itself in a play of definable numbers and forms. The spiral is still our most profound image for the movement of Time and therefore it is central to our vision of evolution. The following passage from Sri Aurobindo's *The Problem of Rebirth* precisely verbalizes what we have just experienced of universal law through the language of geometry:

What is around us is a constant process of unfolding in its universal aspect; the past terms are there, contained in it, fulfilled, overpassed, but in general and in various type still repeated as a support and background; the present terms are there not as an unprofitable recurrence, but in active, pregnant gestation of all that is yet to be unfolded by the spirit: *no irrational decimal recurrence, helplessly repeating forever its figures, but an expanding series of powers of the Infinite.*

This is surely the Will in things which moves, great and deliberate, unhasting, unresting, through whatever cycles, towards a greater and greater *informing* of its own figures with its own infinite reality. (My italics.)

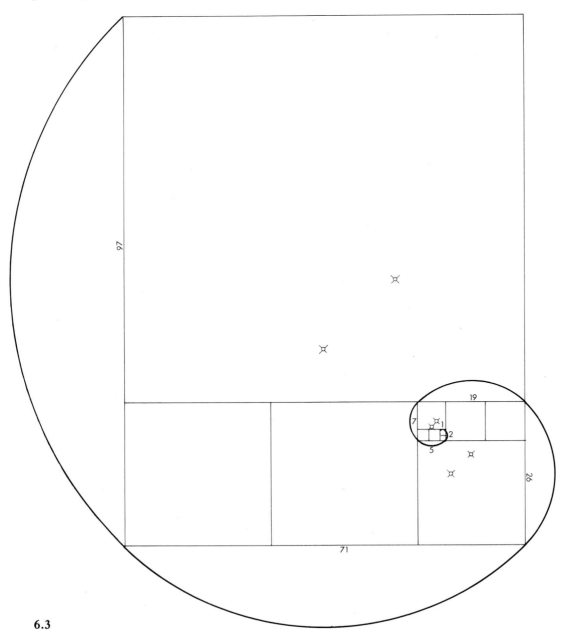

While contemplating the seed head of a shoot-of-pearl grass, we can understand, in view of the above demonstrations on the formation of spirals, the ancient Tantric aphorism, 'Form is the envelope of pulsation.'

The logarithmic spiral is so rich in geometric and algebraic harmonies that traditional geometers named it *Spira mirabilis*, the miraculous spiral. While the radius of this spiral increases in a geometric progression, the radial angle increases in an arithmetic progression. These are the two numerical progressions which yield all the ratios from which the musical scales are constructed. So we can find in these spirallings of gnomonic figures a close association between the temporal laws of sound and the proportional laws of space.

The growth of the human brain seems to have evolved through a gnomonic expansion. The same bulb (the inner or hind-brain) which dominated during the reptilian phase of evolution is still present within us. Above this is the midbrain, limbic area which was the dominant mental apparatus during mammalian evolution; and finally came the emergence of the cerebral cortex in higher man.

Gnomonic expansion in nature makes visible patterns of successive stages of growth. This relates to our notions of time in an interesting way. Ordinarily we conceive of time either as a fleeting directional movement from a dissolving past through an imperceptible present towards an imaginary future, or, mystically, as an all-containing, eternal fullness. The gnomonic principle adds a third description of time. This is time as an expanding growth upon growth, an evolution, one might say, belonging to the conscious energies which transcend their transitory forms and substances. As Chinese wisdom says, 'The whole body of spiritual consciousness progresses without pause; the whole body of material substance suffers decay without intermission.' In such a model, past time remains present as form, and the formation grows through pulsating, rhythmic gnomonic expansion. To remove the most recently accreted layer or compartment from a nautilus shell is actually to move backwards in its lifetime. Logarithmically developed forms always carry this element of the retention of past time, and thereby symbolize the evolution not of substance but of consciousness.

Commentary on Workbook 6

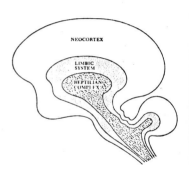

The gnomonic pattern as the basis for the development of the brain through evolution.

句 股 弦 互 求 圖

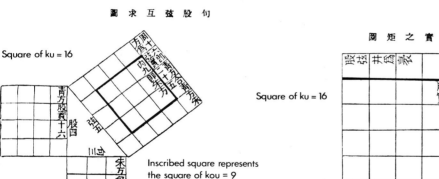

Square of ku = 16

Inscribed square represents the square of kou = 9

square of kou = 9

句 實 之 矩 圖

Square of ku = 16

Ancient Chinese mathematical problems dealing with the gnomonic principle.

71

In gnomonic time, all phases exist in ever-present layers, like the light-year structure of galactic space which makes every glance into a starry night sky a view into the past of distant bodies, while the layers of light just behind the visible light are the future energy waves which will strike and influence the earth. All aspects of the material world, including our own bodies, are therefore in the past tense, existing in a residual gnomonic layer already bypassed by inflowing cosmic energies. This is a rather disturbing idea, yet not dissimilar to our ancestors' notions of time. The *Atharva Veda* says,

> Name and form are in the Residue. The world is the Residue. Indra and Agni are in the Residue. The universe is in the Residue. Heaven and Earth, all Existence is in the Residue. The water, the ocean, the moon and the wind are in the Residue.

In Egyptian iconography the square and its gnomon appear on the throne of Osiris upon which the king takes his seat. The enthroned king, as representative of the eternal solar power on earth, is thus appropriately associated with the fixed element, the square with its gnomon, that which is constant through growth and change. Yet this throne is also the throne of Osiris – the divinity representing the cyclic pattern of change in nature – in his other-worldly kingdom of potentiality. In this sense the throne is the fixed support upon which the Osirian cycles of flux must rest.

The throne upon which Osiris sits is clearly depicted as the square of 4, as it transforms into the square of 5 through the principle of $\sqrt{5}$ on which all the ϕ proportions rest. It is therefore shown as the seat of the world of transformation through death and rebirth, represented by Osiris.

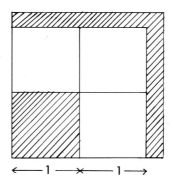
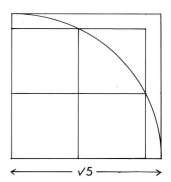

This figure also images the passage from 4 to 5, that is from the elemental or mineral realm associated with the number 4 to the realm of life associated with the number 5, since Nature begins to create pentagonal figures only with the advent of life. The original unity within the four squares of the 2^2 is projected outward to form the gnomon, the fifth part, which is equal in area to each of the other four squares.

The living king is not only the earthly representative of the eternal solar power, but he is also Horus, the son of Osiris, who receives and brings his father's essence-force into the world again. The relation of father to son or of the dead king to the living king was greatly stressed in traditional society and can be seen as the pulsation of the gnomonic retention of the past into the present and future. If the power and influence of the dead king, the father, are related to the original square, and the energies and activities of the living king to its gnomonic expansion, we have an image of a social order based upon the relation of the individual to his ancestral community. The amazing continuity of ancient Egyptian culture over three millennia demonstrates a continual innovation in which it seems that nothing of the essential experience of the past was ever lost.

The square and its gnomon then serve as an archetypal image of certain kinds of growth in nature, and as an image of time and of evolution itself. Such a figure has the value of helping one see beyond the surface of things to identify an underlying pattern, a function with its own dynamics and mechanism.

In the philosophic approach to geometry we are attempting the contemplation of characteristics of form as carrying a meaning in themselves. There is a teleological message contained, for instance, in the spiral itself; for it moves in successively opposite directions towards the ultimate expression of both the infinitely expanded and the infinitely contracted. The spiral is constantly approaching these two incomprehensible aspects of the ultimate reality, and therefore symbolizes a universe moving toward the perfect singularity from which it arose. Thus the spiral-like arms of our galaxy constitute an image of the continuity between fundamental polarities – infinite and finite, macrocosm and microcosm.

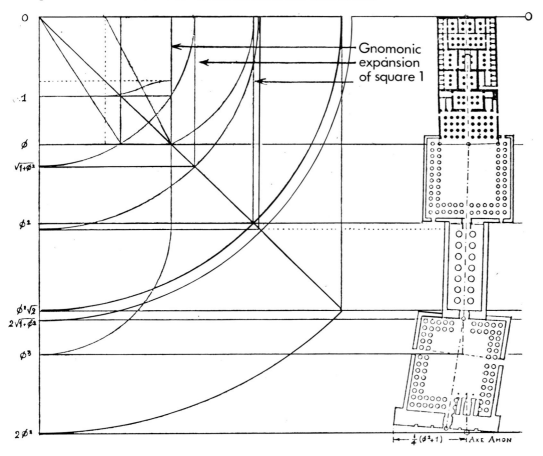

The Egyptian Temple of Luxor uses, like the Hindu temple, the principle of gnomonic growth in its architecture, but in a much different manner. Here the phases of the temple's construction, which are directed by the various ϕ proportions of the initial square of the inner sanctuary, are coincident with the phases of growth of a human body, which the whole of the temple plan symbolizes.

VII The Squaring of the Circle

There are a number of diagrams in the literature of Sacred Geometry all related to the single idea known as the 'Squaring of the Circle'. This is a practice which seeks, with only the usual compass and straight-edge, to construct a square which is virtually equal in *perimeter* to the circumference of a given circle, or which is virtually equal in *area* to the area of a given circle. Because the circle is an incommensurable figure based on π, it is impossible to draw a square more than approximately equal to it. Nevertheless the Squaring of the Circle is of great importance to the geometer-cosmologist because for him the circle represents pure, unmanifest spirit-space, while the square represents the manifest and comprehensible world. When a near-equality is drawn between the circle and square, the infinite is able to express its dimensions or qualities through the finite.

Workbook 7
Squaring the circle

In the following pages I will invite you to follow a squaring of the circle which I feel contains many symbolic keys for the contemplation of universal creation. We begin by drawing a circle, acknowledging it as the geometric metaphor for a homogeneous, non-differentiated space. As in our other diagrams, this unity-space must divide into duality in order to create. So we begin by dividing the unity-circle into two halves, a division which occurs *within* the initial unity.

Drawing 7.1. Draw a circle with centre O and radius $OA = 1$. Draw the diameters AA' and BB', at right angles. With centres on the diameter BB'

draw two circles, each having a radius of half that of the original circle. From point A swing an arc NM which is tangent to the circumferences of the two inner circles. Repeat from point A'. Construct square $ACB'O$ from the radius OA of the original circle.

As shown by the arc of the semi-diagonal of this square, the radius AE of the arc NEM is ϕ, and the arcs NEM and NDM divide the radii AO and $A'O$ into the Golden Division of $1/\phi$ and $1/\phi^2$.

A curious paradox arises when we divide a unitary circle in two in this manner, forming the basis for the traditional yin-yang symbol. The two circumferences of the inner circles are together

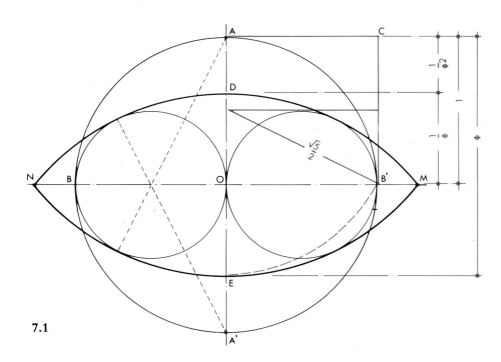

7.1

equal to that of the larger circle, but the area contained within the two is only half that of the original circle. One has become Two. Both Hindu mythology and medieval European alchemy give us the same metaphor to contemplate this mystery of a homogeneous unity which becomes a polarized duality: When homogenized or thoroughly stirred milk is left to stand in a moderate temperature it will enter into an acidic fermentation which coagulates the milk into the contracted, fatty globules of curd which float in the watery whey. We have then a separation of mutually repellent forms which arise from a common source. Mythologically this natural process is figured in Cain and Abel, Seth and Horus, Indra and the Asuras, etc., the universal, antagonistic and oppositional interaction which forms life: this is yin and yang.

When we form geometrically the container of the two circles by drawing an arc from each end of the vertical diameter tangential to the two circles, terminating both the upper and lower arcs at the horizontal diameter, we see that these two arcs have cut the vertical radius OA (regarded as 1 or unity) into the Golden Section of $1/\phi$ and $1/\phi^2$. The Golden Mean as the primary division of Unity is here analogous to the invisible provocator, the universal contractive or coagulating power. It is also evident that the radius of this arc equals $1 + 1/\phi$, which is ϕ.

The vesica enclosing the primary duality (similar to the Vesica Piscis of Workbook 2, but of different proportions), is found everywhere in Egypt as the symbol of Rê, the solar, life-giving force, the emanating Word, the mouth which speaks the names of the gods, the fractions. The mouth of Rê also resembles the moving shape around a vibrating string (see p. 22).

Drawing 7.2. The arc of Rê which lies tangent to the two inner circles cuts the outer unity circle at exactly the point which gives the side of a regular pentagon inscribed in the outer circle, measured from the upper extreme end of the vertical diameter to the left at J and to the right at F. In addition, by placing the compass on the lower extreme of the vertical diameter and drawing an arc tangential to the near curve of the twin circles, we can obtain the exact length of a third side of the same inscribed pentagon touching the outer circle to the left at H and to the right at G. Then, by simply connecting the two upper ends of the pentagon to each end of the base side, we form a perfect inscribed pentagon.

Thus, given with the original scission or contraction into two is the plan of the return: the pentagon, the symbol of life, with its fivefold symmetry which appears only in living organisms. This is the figure ascribed to the physical and vital aspects of man, who, through the five senses, perceives the natural world and thus brings it into existence. The star pentagram formed by the diagonals within the pentagon symbolizes transformed or perfected humanity, because all the line segments of the star pentagram are derived from the Golden Proportion (see p. 52).

Therefore the initial division, which simultaneously gives the proportions for the fivefold symmetry, carries with it a teleological message which is that of Life as the force of levity and return toward the light, as we see in plants which grow back toward the radiant energy source which they embody. This uplifting is given geometrically at the instant creation begins, when One becomes Two. Now that this principle is invoked in our geometric metaphor of creation, we can proceed with the symbolic squaring.

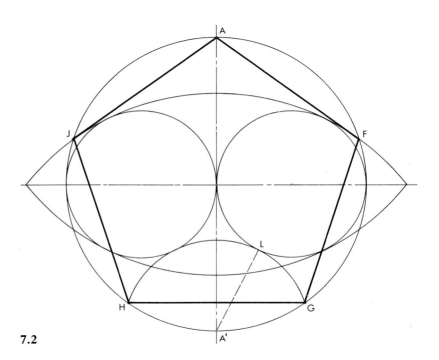

This medieval squaring of the circle through the pentagon is meant to symbolize the harmonization of intuition (indicated by the pentagon) with reason (indicated by the square); or the idea that the infinite (the circle) communicates with the human intelligence through the laws of harmony.

Drawing 7.3. Enclose the initial circle in a square. Then draw a circle by using the centre of the initial circle as centre, and the distance to the tip of the vesica as radius. This circle will be equal in circumference to the perimeter of the square which is tangent to the initial circle.

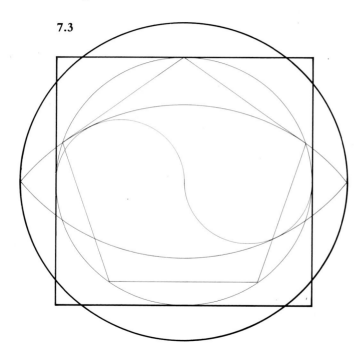

7.3

Drawing 7.4. This is based on the following:
The radius of the circle circumscribing the mouth of Rê, by Pythagoras:

$$\phi^2 = 1 + r^2$$
$$r = \sqrt{\phi - 1}$$
$$r = \sqrt{\phi}$$

and the circumference equals $2\pi\sqrt{\phi}$, with $\sqrt{\phi} = 1\cdot272\ldots$ and

$$\pi = 3\cdot142\ldots$$
$$2\pi\sqrt{\phi} = 7\cdot993 \text{ for the circumference, or}$$
approximately 8.

We know that the square circumscribing the original circle whose radius is 1, has a side of 2. Thus the perimeter of this square is 8, and therefore approximately equal to the circumference of the large circle, 7·993.

This leads to the value of π which is believed to have been used by the ancient Egyptians for the construction of the Great Pyramid:

$$2\pi\sqrt{\phi} = 8$$
$$\pi\sqrt{\phi} = 4$$

then,

$$\sqrt{\phi} = \frac{4}{\pi} = 1\cdot272\ldots$$
$$4\sqrt{\phi} = \pi = 3\cdot1446056\ldots$$

Whereas true π is 3·1415926 . . . . A nearly exact π using the Golden Mean is $\phi^2 \times 6/5 = 3\cdot1416404\ldots$. The ratio 5:6 or 1:1·2, incidentally, is the function which relates ϕ to π, and 1·2 equals the relationship of 12 to 10. Twelve is the number of the circles of cosmic time, it is the number of completion, and as the ratio 6 to 5 it relates the hexagon to the pentagon.

To return to our figure, by using the side of one quarter of the square (which is identical to the radius of the first circle) as Unity, we can determine these values:

$$pn = \frac{\sqrt{5}}{2} = 1\cdot118\ldots = \frac{1}{2} + \frac{1}{\phi}$$

$$B'n = B'K = A'M = \phi = 1\cdot618\ldots$$

$$OD = On = \frac{1}{\phi} = 0\cdot618\ldots$$

$$AD = \frac{1}{\phi^2} = 0\cdot3819\ldots$$

$$OM = \sqrt{\phi} = 1\cdot2720196\ldots$$

$$AF, HG = \sqrt{(1 + 1/\phi^2)} = 1\cdot1756 = \text{side of pentagon}$$

$$DM = \sqrt{2} = 1\cdot4142135\ldots$$

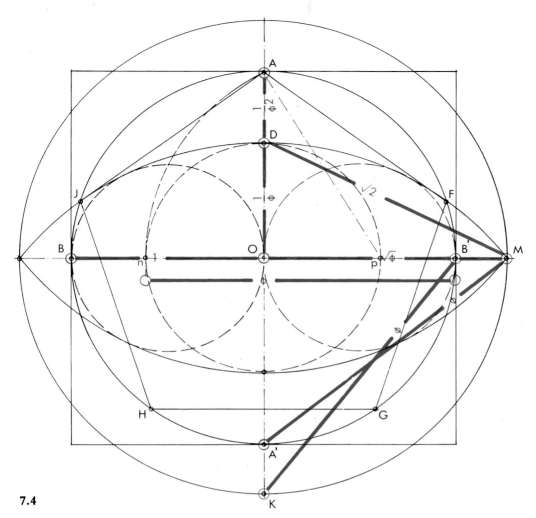

7.4

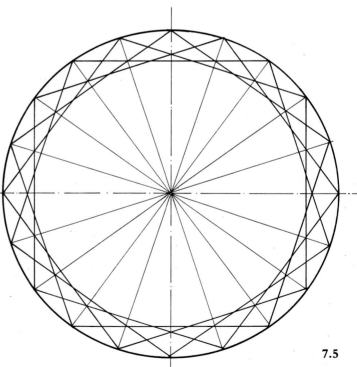

Drawing 7.5. The next objective is to construct a square equal in area to the original circle. To inscribe three additional pentagons within the circle, bisect one side of the pentagon, mark the corresponding point on the circle, then bisect the resulting segments. This gives starting points for the three new pentagons, so that the total of apexes is 20.

This can symbolize for us the quintessential five-fold symmetry, the flowering of the life principle in its return towards light, expressing itself in terms of the fourfold symmetry of elemental nature, earth, air, fire, water.

7.5

Drawings 7.6, 7.7, 7.7a. If we begin at point *A* where the first pentagon touches the vertical axis, and draw a straight line through the second and fifth points of the pentagon vertices, then extend these lines to the vertical and horizontal axes (*PO*), this will be the first side of a square. Continue drawing this to form lines *QR*, *RS* and *SP*. By using the geometric methods of calculation on the pentagon and its diagonal from Workbook 5 we can determine the values given in Drawings 7.7 and 7.7a and thus verify that this new square will be approximately equal in area to the surface of the initial circle. Half the diagonal of the square, *OP* = 1·26006, and the side of the square *PQRS* = 1·26006 × $\sqrt{2}$ = 1·7819938.

This is a squaring taken from a design of the Middle Ages and is not mathematically very exact, but symbolically it has great simplicity and beauty. The numbers given will show the side to be 1·7819938 while a more perfect square would be 1·7724397, making a difference of 0·0095548, or a π equal to 3·17.

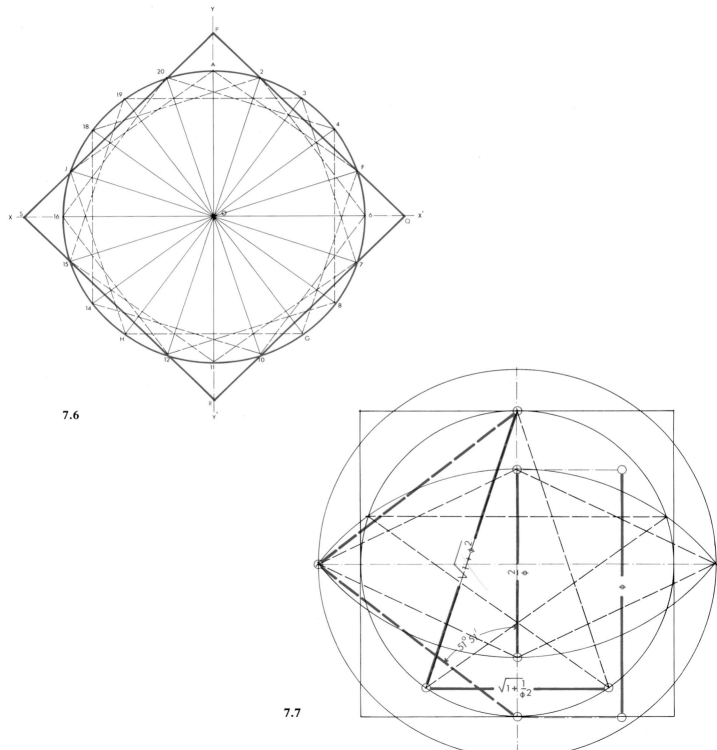

7.6

7.7

The circumambulation of the Ka-ba (cube) at Mecca is a symbolic ritual related to the concept of the Squaring of the Circle.

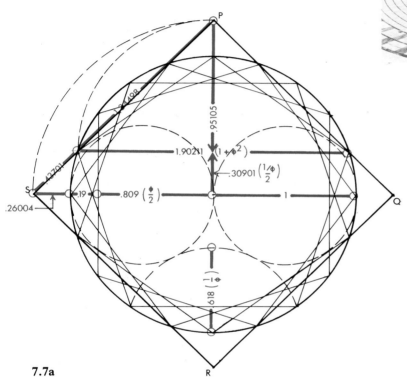

7.7a

Drawing 7.8. In combining the drawings we notice that the vesica or mouth of Rê which makes the initial abstract (linear) square does *not* touch, but rather 'emits' the second manifest squaring (the one of surface). Here we have in one diagram the classical geometric relationship made between the circle and square, between the spiritual and material worlds. In the next section we will discuss this same relation in volume, between the sphere and the cube.

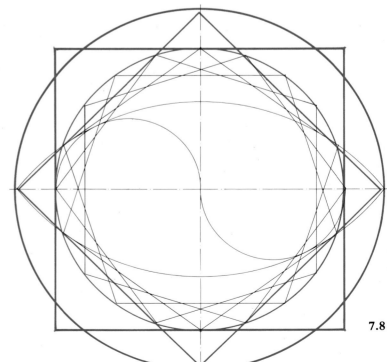

7.8

VIII Mediation: Geometry Becomes Music

We have been considering the division of unity through both the idea of the root function (the generative root of 2 and the regenerative root of 5) as well as through the idea of the three- and four-term proportions which result from them. In this section we will bring the ideas of proportion and roots together so that their relationship can be thoroughly understood, and at the same time show how this resulting geometry becomes the basis for musical harmony. Hopefully this will shed light on the statement of Goethe, 'Geometry is frozen music'.

The best approach to these objectives is by way of what is considered the cornerstone of ancient philosophical mathematics, the *science of mediation*, which is a simple observation of the functions of mean terms. Using our discussion on the three- and four-term proportions (p. 44) as a point of departure, let us first take heed of Plato's warning that comparisons based on four elements, that is 'discontinuous, four-term proportions' are of the nature of what he calls 'particular knowledge', which is of a vulnerable character, open to dispute and arbitrariness. Opposed to this is what he calls 'essential knowledge', which is not the simple accumulation of factual or even conceptual data pertaining to objects or phenomena, but rather consists of an awareness of the metaphysical constructs through which the mind is able to gain its comprehension. The laws which govern the creation of things are the same laws as those which allow for their comprehension, and essential knowledge is an understanding of these laws. Such knowledge can be gained, Plato says, through the study of mediation, which is the binding of two extremes by a single mean term. We have seen the example of this involving ratios made up of three terms, $a:b::b:c$, what we call the geometric proportion and the Greeks called *logos*. But this simple example is not the only three-term proportion, and the science of mediation explores all the proportional statements possible between three terms not only through a direct proportional relationship but also through the play of *difference*.

A mediating proportion can be defined as *A group of three unequal numbers such that two of their differences are to each other in the same relationship as one of these numbers is to itself or to one of the other two numbers.*

This strange mathematical 'koan' contains the formula for the three major medians, the Arithmetic, the Geometric and the Harmonic.

Let us go through this definition of the three medians step by step. A median proportion is formed from a group of any three numbers with a greater than b and b greater than c $(a > b > c)$, such that '. . . two of their differences', that is:

$a - b$ (this is one difference)
and $b - c$ (this is the second difference)

'. . . are to each other'

$a - b : b - c$

'. . . in the same way as one of these numbers is to itself' (case 1):

$a - b : b - c :: a : a,\ b : b,\ c : c$

'. . . or as one of these numbers is to one of the other numbers':

(case 2) $a - b : b - c :: a : b$ or
(case 3) $a - b : b - c :: a : c.$

In case 1 the expression, when solved for the mean term b, becomes $b = (a+c)/2$, which is the general formula for an arithmetic proportion. 3, 5, 7 is an arithmetic progression with the arithmetic mean, $b = 5$.

In case 2 the expression, when solved for the mean term b, becomes $b^2 = ac$ or $b = \sqrt{ac}$ which is the general formula for the geometric proportion. 4, 8, 16 is a geometric progression with the geometric mean, $b = 8$.

In case 3 the mean term b is then $b = 2ac/(a+c)$, and this is the general formula for the harmonic proportion. 2, 3, 6 is a harmonic progression with the harmonic mean $b = 3$.

This statement of mediation gives us then the general formula for all our basic mathematical operations. The arithmetic proportion contains the law for addition and its inverse, subtraction, and describes the relationship which gives the natural series of cardinal numbers, 1, 2, 3, 4, 5, 6, . . . etc. The geometric proportion contains the law for multiplication and its inversion, division, and describes the relationship which gives any series of geometric progressions. As we have said, addition and multiplication are mathematical symbols for patterns of growth. The harmonic mean is derived from a combination of the first two; it is formed by a multiplication of any two extremes (a, c) followed by the division of this product by their average or arithmetic mean $(a + c)/2$. For example, given two extremes, 6 and 12, the product of 6 and $12 = 72$, the arithmetic mean between 6 and 12 is 9, and $72 \div 9 = 8$, so 6, 8, 12 is an harmonic proportion.

$$\text{Arithmetic:} \quad b = \frac{a+c}{2}$$

$$\text{Geometric:} \quad b^2 = ac$$

$$\text{Harmonic:} \quad b = \frac{2ac}{a+c}$$

Each proportion has a number of characteristics which are peculiar to it. For example, the arithmetic proportion shows an equality of difference, but an inequality of ratio. Thus, in the arithmetic proportion: 3, 5, 7,

$$7-5 = 5-3 \text{ but } 7/5 \text{ does not equal } 5/3.$$

A geometric proportion on the other hand, is characterized by an equality of ratio but an inequality of difference. Therefore in the geometric proportion: 2, 4, 8,

$$4/2 = 8/4 \text{ but } 4-2 = 2 \text{ does not equal } 8-4 = 4.$$

The most important and mysterious character of the harmonic proportion is the fact that the inverse of every harmonic progression is an arithmetic progression. Thus 2, 3, 4, 5, is an ascending arithmetic progression while the inverse series, 1/2, 1/3, 1/4, 1/5 is a descending harmonic progression. In music it is the insertion of the harmonic *and* arithmetic means between the two extremes in double ratios – representing the octave double – which gives us the progression known as the 'musical' proportion, that is 1, 4/3, 3/2, 2. In other words, the arithmetic and harmonic means between the double geometric ratios are the numerical ratios which correspond to the tonal intervals of the perfect fourth and the perfect fifth, the basic consonances in nearly all musical scales.

The basic proportional structure which contains the axioms for our primary mathematical operations is also the basic proportional structure for the laws of music. Let us then investigate further the role of these three proportions as the archetypal thought forms for the entire universe of music.

The progression 1, 4/3, 3/2, 2 represents the frequencies of a fundamental, fourth, fifth and octave. We then find the arithmetic and harmonic proportions between the string lengths 1 and 1/2 representing the division of the vibrating string in half

The musical octave is based on a tone whose vibrational frequency is in an exact ratio of 2:1 with another tone. On the guitar, for example, if we pluck the whole first string, *EX*, we will sound a fundamental tone called in musical notation E. For ease of calculation let us give this sound the value of 6, designating its vibrations per second (actually 82·5). If we then hold our finger on the fret marked *E'* and then pluck the string length *E'X*, which is exactly half the length of *EX*, its frequency of vibration will be double that of *EX*. It is thus given the numerical value of 12, forming the 2:1 ratio with 6. The tone *E'X* = 12 is called the octave of E. An octave sound has the strange characteristic that it is of the same quality as the fundamental tone, so much so that it seems to blend into it, yet it is definitely higher in pitch. The experience of hearing the octave contains the mystery of a simultaneous sameness and difference. This quality of perceiving both sameness and difference is part of the poise of mind that sacred geometry means to cultivate: one which is precisely discerning yet harmoniously integrating.

Likewise if we place our finger on the guitar fret marked *B* and sound the string length *BX* the tone will be in relation to the fundamental *EX* as 3:2, or as we've shown, 9:6. This tone B is a beautifully consonant sound and is called the musical fifth because it is the fifth tone in a natural series of divisions of string *EX*, the diatonic major scale, with E as Do and B as Soh. There is a scale of eight such natural tonal divisions from *E* to *E'*, hence the name 'octave'. If we place our finger on the fret marked A and sound string *AX* it will sound another consonant note called the fourth, and its frequency will be in relationship to the fundamental as 4:3, or as marked here, 8:6.

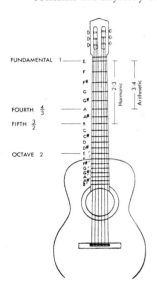

FUNDAMENTAL 1 — E

FOURTH 4/3 — A

FIFTH 3/2 — B

OCTAVE 2 — E'

2·3 Harmonic

3·4 Arithmetic

which produces the octave increase of frequency. This gives the progression 1, 3/4, 2/3, 1/2, because the harmonic mean between 1 and 1/2 = 2/3, the musical fifth, and the arithmetic mean between 1 and 1/2 = 3/4, the musical fourth. In comparing these two progressions, we see an inversion of ratios and a crossing of functional positions between the arithmetic and harmonic mean.

The mystery of musical harmony which develops out of a simultaneous inversion also contains a simultaneity of addition and multiplication. The octave of a fundamental is achieved by the addition of the intervals: in string lengths the fifth plus the fourth equals the octave, and also the multiplication of the vibrational frequencies of the fourth and the fifth equals the octave ($4/3 \times 3/2 = 2$). The combined effect of addition and multiplication produces the logarithm in mathematics and, as we have seen, the Golden Proportion is the archetype for this form of growth.

The above table expresses the explicit mystery of the law of sound, which is that that numbers considered as frequency ratios in a rising scale are equal to the string lengths for the descending scale. The law of musical harmony, when viewed from the idea of mediating proportion, becomes a symbol for the law of natural order, the *Tao* of the created worlds, where oppositional yet simultaneous movements interact to create both sound and form.

We can now begin to visualize this numerical and harmonic principle as geometry.

The geometric mean is found by the formula $b^2 = ac$;

The harmonic mean answers to the formula $b(a+c) = 2ac$; that is the product of the sum of the extremes, multiplied by the mean is equal to two times the product of the extremes, or

$$b = \frac{2ac}{a+c}$$

The geometric proportion is called the *perfect* proportion because it is a direct *proportional* relationship, an equality of proportion bound by one mean term. The arithmetic and harmonic medians work out this perfection through an interchange of differences in a play of alternation and inversion.

	NOTE	H M (FOURTH)	A M (FIFTH)	OCTAVE
VIBRATION	1	4/3	3/2	2
	6	8	9	12
STRING LEGNTH	12	9	8	6
	1	3/4	2/3	1/2
	NOTE	A M (FOURTH)	H M (FIFTH)	OCTAVE

This table shows the simultaneous inversing and crossing of the arithmetic and harmonic mean terms in the musical proportion, as considered from the point of view of vibration and string length.

Workbook 8
Geometry and music

Let us now try to find verifications in number progressions for what I have just stated in words. Taking the geometric series first, we line up two geometric series (of rate 2), one starting with the first odd (male) number after unity, 3, and the other starting with the first even (female) number, 2. 1:2 numerically symbolizes the octave, the spatial milieu in which the first consonant division by 3 (giving the fifth 2/3) symbolizes the seeding, form-giving function which enters and specifies the fixed proportional divisions within the primal ocean of the undifferentiated sound, the octave.

3   6   12   24   48

2   4   8   16   32

In the *Timaeus* Plato demonstrates that the multiplication of 2 and of 3 gives us all the numbers for the Pythagorean tuning system by successive multiplication by fifths (3:2). And as Platonists we remember that Two symbolizes the power of multiplicity, the octave, the female, mutable receptacle, while Three symbolizes the male, specifying, fixing, immutable pattern-giver whose multiplication table gives the entirety of music. This was the 'music of the spheres', the universal harmonies played out between these two primal male and female symbols.

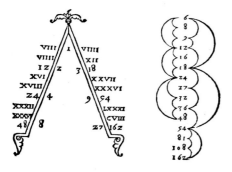

This diagram by Giorgi shows the two progressions of 2 and 3, as given by Plato in the *Timaeus*, placed in association with the musical proportion of 6, 8, 9, 12. It uses the musical proportion as a basis for generating number for a succession of musical octaves, fourths and fifths, thus constructing an harmonic system which could be used as a model for architecture, painting, and other arts.

Now let's interpenetrate these two geometric series so that the geometric progressions act as a sort of copulation:

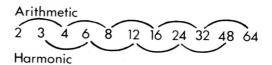

We can see here that every other overlapping set of three numbers provides us with alternately an arithmetic and an harmonic proportion: 2, 3, 4 is arithmetic; 3, 4, 6 is harmonic; 4, 6, 8 is arithmetic; 6, 8, 12 is harmonic, etc. So the interfusion of the male number, geometrically generated, with the female number, also geometrically generated, provides us with these two alternating proportional possibilities.

Now let us take the same thing we have seen in a linear structure and look at it in a formal structure, through the Table of Lambda:

1	2	4	8	16	32	64
	3	6	12	24	48	96
		9	18	36	72	144
			27	54	108	216
				81	162	324
					243	486
						729

This is a triangular array of numbers which crosses the geometric progression by 2 (horizontal) with the progression by 3 (diagonal). All the successive vertical numbers are to each other in the ratio of 2:3, which is the same as multiplying one term by 3/2 in order to obtain the term below. This successive multiplication by 3/2, the musical fifth, is the method used by the Pythagoreans for generating the musical scale. The origins of the number series which appears on pp. 82 and 83 will now be evident.

The generative character of the Lambda Table is emphasized in the woodcut of 1503 on p. 7 by its portrayal on a woman's thighs. In examining the table we can see that each square of four numbers, for instance 2, 4, 6, 3, contains within it two arithmetic progressions (that is 2, 3, 4, and 2, 4, 6) giving us three sides forming the top of a square and one diagonal. We see in the same figure the harmonic progressions 2, 3, 6 and 3, 4, 6 giving

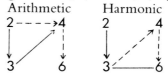

three sides of a square, two of them overlapping with the first proportion, the other giving the fourth side of the square and the other diagonal. So we have in this Table of Lambda, handed down to us by Nicomachus of Gerasa, an intercopulation of these two proportions producing the *square*, which as we have seen, is the symbol of the finite, knowable, manifested realms. These are the numbers and musical proportions from which Plato said the World Soul was fashioned.

Another geometric exercise shows the relationship between the root functions and the mediating principles which create the world of harmony in music.

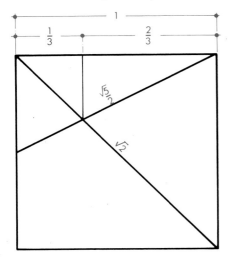

Drawing 8.1. Using the square as unity with both its side and area equal to 1, we see by geometric evidence or by trigonometry that by crossing the $\sqrt{2}$ with the $\sqrt{5}/2$ and simply leading a perpendicular from the point of intersection up to the side (1), we divide the unity into 1/3 and 2/3, and using the unity as the largest term, we have a three-term arithmetic proportion: 1/3, 2/3, 1.

Drawing 8.1. $\dfrac{1}{3} : \dfrac{2}{3} : 1$ Arithmetic

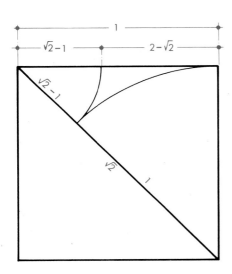

Drawing 8.2. Again using the square for unity and by means of an arc from the lower left corner, bring the length of side 1 down to intersect with the diagonal $\sqrt{2}$. Then lead an arc from the upper right corner back up to the upper side of the square. We have again a point on the upper side at which to divide the square, but this division creates a three-term harmonic proportion, $(\sqrt{2}-1)$, $(2-\sqrt{2})$, 1.

Drawing 8.2. $(\sqrt{2}-1) : (2-\sqrt{2}) : 1$ Harmonic

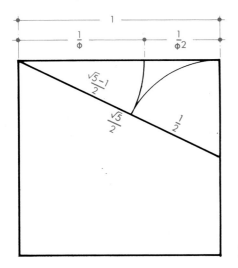

Drawing 8.3. The last division of the side of square 1 is accomplished with the $\sqrt{5}/2$. This is done by leading an arc from the point of intersection of the half-side with the semi-diagonal and the arc with radius equal to the half side to the upper side of the square. This divides our unity into a geometric proportion, $1/\phi^2 : 1/\phi : 1$.

Drawing 8.3. $\dfrac{1}{\phi^2} : \dfrac{1}{\phi} : 1$ Geometric

	Large	Middle	Small
Arithmetic mean	1	$\frac{2}{3}$ (0·666)	$\frac{1}{3}$ (0·333)
Harmonic mean	1	$2-\sqrt{2}$ (0·586)	$\sqrt{2}-1$ (0·414)
Geometric mean	1	$\frac{1}{\phi}\ \frac{\sqrt{5}-1}{2}$	$\frac{1}{\phi^2}\ \frac{3-\sqrt{5}}{2}$

In given square *ABCD* with side 1 draw its diagonals *AC* and *BD*. With radius *BD* and centre *B* swing arc *DC* to form $BG = \sqrt{2}$. With radius *CG* and centre *C* swing arc *GF*. With radius *AF* and centre *A* swing arc *FB* to complete half of the profile of the 'grail'. Repeat on opposite side to complete the figure.

This is the *analogos* or geometric proportion as it is expressed in the division into extreme and mean terms, but within the initial unity itself.

All three medians have been constructed under the condition that 1 is the largest of the three terms. This series was considered as a configuration of transcendental (supra-rational) proportions, as they are all incommensurables contained within the initial Unity. (Remember that ancient music itself is constructed from whole number ratios only, but the principle of musical structure belongs to the supra-rational divisions of Unity.) The three medians comprise the trinity of trinities, three unique proportional expressions each of three terms. They express through the sacred roots of 2 and 5 the essential harmonic division of both Time (music) and Space (geometry) and have often been employed in traditional cultures for the bases of architecture, art, science, mythology and philosophy.

Drawing 8.4. Here is a means of drawing a beautifully proportioned cup or grail-shaped vessel, using only the harmonic division to establish its curves and measures. We may speculate that this is the geometric essence of the Holy Grail.

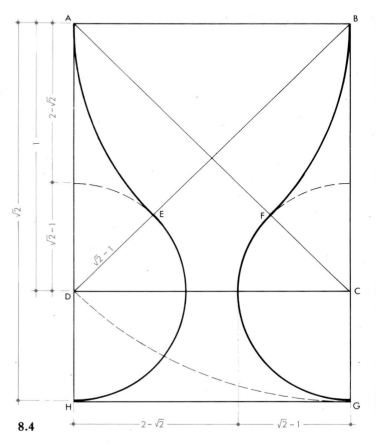

8.4

It is perhaps because the contemplation of the laws of mediation allows one to glimpse the fundamental relationship between Music and Geometry that Plato, in his *Seventh Letter*, says that it is more revered than any other study of knowledge. And perhaps for the same reason the Egyptians constructed two great pyramids at Giza, one of which is based on 1, $\sqrt{\phi}$, ϕ, the only triangle whose sides are in the geometrical progression, and the other whose sides are in the arithmetical progression 3, 4, 5. In our day Simone Weil speaks of the importance of this study as the philosophical basis for Christian mysticism.

It is in the work of Hans Jenny that we can begin to see the relationship of form and sound in the physical world. Jenny's experiments have shown that sound frequencies have the propensity to call into arrangement random, suspended particles, or to organize emulsions in hydro-dynamic dispersion into orderly, formal, periodic patterns. In other words, sound is an instrument through which temporal frequency patterns can become formal spatial and geometric patterns.

Commentary on Workbook 8

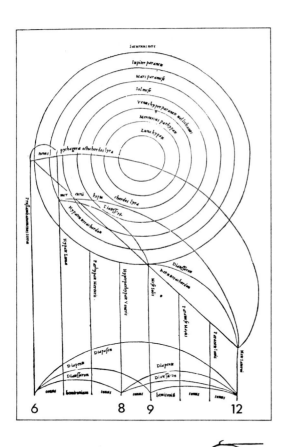

A planetary system based on the musical proportion, 6, 8, 9, 12, of the arithmetic mean and the harmonic mean between the geometric ratios of 6 and 12, along with the other tones of the Pythagorean diatonic (major) scale.

Albrecht Dürer's human canon is entirely composed of proportions derived from the three unique divisions of Unity into the Arithmetic, Harmonic and Geometric Proportions.

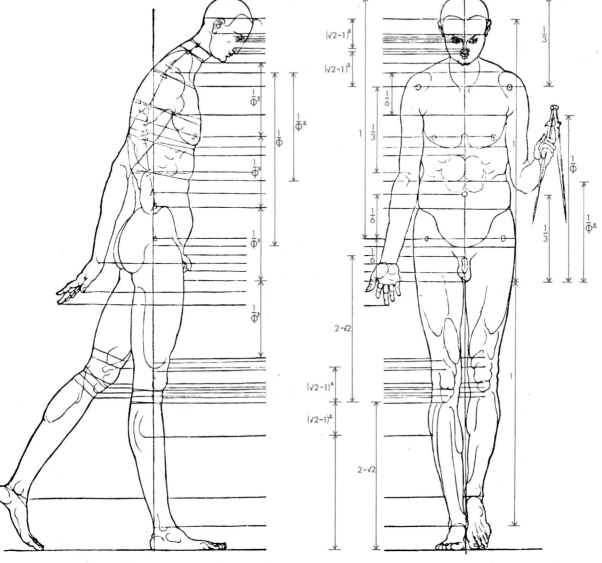

Sound frequencies in this experiment cause random particles to assume geometric patterns.

RIGHT Geometric symmetries are generated from electronic beam frequency interference patterns. The sevenfold figure arises out of the circle and returns to it.

Vibrational image of the Seed Sound, Om.

This drawing combines two important figures, the 3, 4, 5 triangle and the Golden Mean, to produce the musical ratios. Although we give the steps to construct the diagram, it is not recommended for beginning geometers. It is based on a drawing from *The Divine Proportion* by H.E.Huntley.

Draw a circle with centre L and radius LA and draw diameter AC. Draw a line perpendicular to AC and passing through A. Swing an arc with centre A and radius AC to F. Draw a line from F through the centre L to intersect the circle at H. Draw a line perpendicular to FH from A to insersect circle at D.

Repeat with CB perpendicular to FH. Draw rectangle $ABCD$. ($ABCD$ is a 1:2 rectangle, proof being that $LA = \frac{1}{2}AF$. Triangle LJC is similar to triangle LAF. $JL = \frac{1}{2}JC$. $BA = \frac{1}{2}BC$.)

Construct 3, 4, 5 triangle by drawing a line from F tangent to the circle at D and extending it to intersect the diameter AC at E. Proof of the 3, 4, 5 triangle is made by the Egyptian method of angular addition: AFM and DFM are both 1:2 angles.

$$\text{angle } \frac{1}{2} + \text{angle } \frac{1}{2} = \frac{(1\times2)+(1\times2)}{(2\times2)-(1\times1)} = \frac{2+2}{4-1} = \frac{4}{3} \quad \frac{AE}{AF} = \frac{4}{3}$$

Draw a circle with centre J and radius JN.

Relationships with short side of rectangle $(AB) = $ Unity:

$ML = \frac{1}{2}$

$HM = JN = GJ = \phi$ (by rotation of semi-diagonal LC about L to H)

$GM = JG + JM = \phi + 1 = \phi^2$

$MN = MK = JN - JM = \phi - 1 = 1/\phi$

$KJ = JM - MK = 1 - 1/\phi = 1/\phi^2$

$GK = GJ + JK = \phi + 1/\phi^2 = 2$

$JQ = 1/\phi$ (by similar triangles JKQ and MKA)

$JR = JG = \phi$

If unity is considered to be one unit of the 3, 4, 5 triangle, instead of the side of the 1:2 rectangle, a second series of ϕ relationships is engendered along with a series of whole number ratios which are fundamental to the formation of the musical scales:

If $AF = 3$, instead of $\sqrt{5}$, and Unity $= AF/3$ then the side of the 1:2 rectangle $= AF/AC = 3/\sqrt{5} = 1\cdot3416$.

Then $RF = 3\times\phi = 4\cdot854$.

$HF = 1/\phi \times 3 = 3/\phi$

$HN = AF = 3$

$HF:HN::1:\phi$

$NF:HN::1:1/\phi$

$ED:DF::3:5$

$AT:AE::5:8$

$AL:AF::1:2$

$AL:AE::3:8$

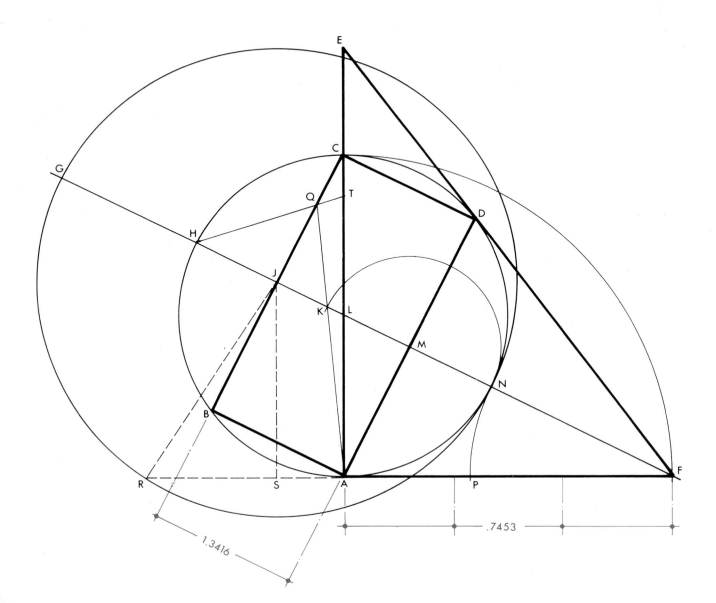

In ancient Egypt the audial sense – that is the direct response to the proportional laws of sound and form – was considered as the epistemological basis for philosophy and science. This is evoked by the blind harpist, whose proverbial wisdom comes not from the visual world of appearance but from an inner vision of metaphysical law.

A number of contemporary sciences are now verifying the ancient cosmogonic vision of a vibratory creation through the Creative Word or Cosmic Sound. Alain Daniélou points out that it was the absorption in this mysterious interchange between vibration and form which was the basis of the great spiritual cultures of the past:

> From the atoms up to the universe, each of the cosmic movements possesses a tempo, a rhythm, a periodicity and can be compared then to vibration, therefore to a sound which expresses its nature. Not all vibrations are perceptible to our ears, but the relationships between the vibrations can be compared to the relations of audible frequencies. All atoms can thus be considered as the forms of an energy which expresses itself in a rhythm, and all substances are characterized by a particular relationship of rhythms which can be represented by a relationship of sounds. It is because of this similarity between the relationships of the sounds on one hand and the forms and substances of nature on the other, that language and music are possible.
>
> The pure sounds, the immaterial sounds which constitute the profound nature of things and which Kabir calls 'their inaudible music', can be perceived through instruments more subtle than our ears. To arrive at their perception constitutes one of the goals of the practice of these curious physio-mental disciplines called *yoga*.
>
> (*Traité de musicologie comparée.*)

And Sir John Woodroffe, speaking from his translations of Hindu texts, says,

> The natural name of a being is the sound produced by the concordant action of the moving forces which constitute it. This is why it is said that he who mentally or physically pronounces the natural name of a being gives existence to the being who bears this name.
>
> (*Garland of Letters.*)

IX Anthropos

The geometric cosmology we have been surveying is part of a mystic doctrine of creation known as anthropocosmic, a doctrine which is fundamental to the esoteric tradition in philosophy since the earliest times, and which has been restated in our time by Rudolf Steiner, R.A. Schwaller de Lubicz and others. The first principle of this theory is that Man is not a mere constituent part of this universe, but rather he is both the final summarizing product of evolution and the original seed potential out of which the universe germinated. We may use the analogy of the seed and the tree: the tree of the universe is the actualization of the seed potential which is Cosmic Man. I am using the word Man here in relation to its Sanskrit root *manas*, meaning 'mind', or the consciousness which can reflect upon itself.

This same image of the identity between seed and tree, or between Cosmic Man and transitory man on the tree of evolution, is given in the Book of Genesis. To elaborate I use some ideas from *The Cipher of Genesis* by the Kabbalistic author Carlo Suarès, putting them in terms of anthropocosmic thought.

In Chapter 1 of Genesis, Adam is placed in the garden with all the animals and plants already created. Adam is the summation or final stage of the evolutionary process. This conforms to the paradigm of Man the container or recapitualizer of the entire evolutionary unfolding which preceded him.

In Chapter 2, Adam (now conceived as the schematic organization of the entire cosmic metabolism) is the first-born thing. In this chapter, which seemingly contradicts the first, YHVH-Elohim creates all the animals and shows them to Adam, and Adam is tested by having to name them one by one. In this test Adam recognizes each species as an offshoot of his own central trajectory. He can name them because he knows them to be of himself. Adam is the core trunk of the evolutionary tree. The animal species are the relatively fixed, specialized lateral branches from the surging core.

The apparent contradiction between Chapters 1 and 2 of Genesis finds a parallel in contemporary embryology, which also gives us two contradictory theories for human development: the 'recapitulation' theory, and that of 'neoteny'. The former, which corresponds to Genesis 1, is the theory that animals repeat the adult stages of their ancestors during embryonic and post-natal growth. Therefore the human embryo passes through all the major evolutionary phases which have preceded him: not only the mammal, reptile, fish and vegetal, but also, in the early stages of cellular division, all the regular geometric solids. Neoteny, however, poses an almost opposite view which corresponds to Genesis 2. This theory is based on the fact that there are over twenty important bodily characteristics which are common to both man and primate, but in the primate they appear in a stage of the embryo or the juvenile and are then outgrown. Physically humans appear as prematurely born primates in which these physical features have been hormonally braked or arrested.

Adam, in naming the various species, recognizes, or shall we say, remembers, his own embryonic past (recapitulation). But he also recognizes himself as the fiery seed, the primal pattern for the total organic process of universal life (neoteny). Adam, at this moment of creation, might declare, 'I see nothing which is not me; I see nothing which is altogether like me.' Thus Adam passes the test. He goes beyond the identification of himself with the successive mineral, plant and animal phases of

The idea of Cosmic Man is echoed in contemporary science in the concept of the hologram, which demonstrates that each fragment of a whole contains the constituents of the overall structure of the whole. At the same time, as a particular partial of that whole, it expresses itself as an individual. In ancient science the metaphoric application of the notion of the Anthropocosm was the basis for astrological philosophy, and in alchemy it may be found again as the search for the Philosophers' Stone – 'that part in which the whole may be found'. In this Renaissance drawing the body of man is placed in relation with the important proportions of universal geometric forms and numerical ratios. Here we see the correlation between $\sqrt{2}$ and the procreative organ of man.

evolution, and at the same time identifies himself with the highest power in the organization of cosmic energy, the unmanifest geometry of the seed-idea. With his identification with his original universal nature, Adam is ready for his incarnation as Adam Kadmon, the embodiment of Cosmic or Divine Man.

The Vedic tradition transmits the same anthropocosmic vision from a more metaphysical position. It tells us that God created the universe from a desire to see himself and to adore himself. The being of this inconceivable God may be considered as an all-conscious, all-containing, all-powerful, homogeneous, endless expanse of pure, formless spirit. His desire to see himself created (or distinguished from himself) an Idea of himself, called in Indian thought the Real-Idea. This divine self-apperception, the 'Creative Word' in Judaeo-Christian thought, *this event itself is Cosmic Man*. And this Cosmic Man is what we, actual man, call the Universe.

The created universe is then seen as a nourishing placenta through which this Divine self-idea is embodying itself or incarnating: a genesis, clothing itself in matter so as to become perceptible and adorable. This position is the inverse of our ordinary thinking. Humanity is not seen as the child or product of Mother Earth, but rather earth is an essential quality contained within the character of Cosmic Man.

Anthropocosmic philosophy images evolution as a continual inversing exchange between eternal Cosmic Man and evolving humanity. The Universal Being performs an involution into the dense seed-form of itself. In principle this is imaged by the mineral kingdom, the extreme of inconscient, fixed densification. This involved seed then provokes an opposite movement of evolution. The plant kingdom follows which then lifts upward and outward; it animates, frees and embodies the divine qualities which were locked or involved in the mineral.

These divine qualities manifested and clarified themselves as functional principles and stages of growth in the plant kingdom – that is root, stem, leaf, flower, fruit, seed – which we can read as symbol-analogies for the entire universal process of Becoming.

The animal kingdom then appears as an inversion of the plant process, and we can detect here an alternating rhythm of involution and evolution which gives rise to the succession of the kingdoms. The animal 'involves' again the principles and activities and vital functions which the plant had 'evolved' or opened, clarified and uplifted. The animal achieves through this involution the power of individual mobility which is the necessary predecessor of individual will. The involution can be considered as the materialization of spirit, and the evolution as the spiritualization of matter.

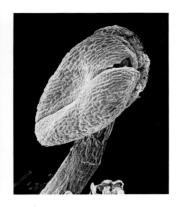

These electron microscope photographs reveal the morphic rapport between flowering or leafing-out processes in plants and the sexual aspects of animal development.

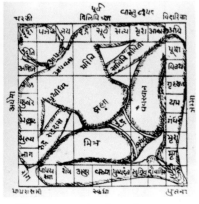

The gnomonic plan of a Hindu temple is superimposed on a diagram of the Purusha or Cosmic Man. The ancient Hindu architectural sutra says 'The universe is present in the Temple in the form of proportion.'

Rudolf Steiner gives an effective image for this process by observing that man in his animal body is in reality no other than a plant turned inside-out. The respiratory function in the plant is the leaf. This function is performed open to the sun at the outer extreme of the principle of branching. In man the respiratory function is the lung; its branches are within.

To continue the analogy, we observe that the flower, which is the sexual organ of the plant, grows upward and pulls the energy of the plant upward toward the light; whereas in man and animal the sexual organs face downward and pull the energies of the body downward. The plant roots in the earth; in man, the characteristic root function is found in the convolutions of the brain which is rooted in the sky of thought and mental energies. The mental process is a process of digestion, assimilation and transmutation which functions as a higher frequency of the intestinal, digestive process, the form of the intestines being also convoluted. In this way the succession of kingdoms from mineral to plant to animal in the physical world become a symbol for the constant movement in involution and evolution of one Being which has divided into the complementary qualities of spirit and matter.

Within the logic of this vision of evolution the purpose of physical man is to transform his involved, animal incarnation into a body of light, just as the plant evolution did for the involved mineral kingdom. Through the vision of Man as Cosmos, the Anthropocosm, sacred geometry becomes a cosmogram depicting the drama of this divine birth. And during all temple-building epochs the sacred architecture based on this geometry has been a book revealing this eternal drama.

In India the *Vastupurushamandala*, the tradition of temple design founded on Cosmic Man, is still alive. We also find that the architectural model for the great Gothic cathedrals was the universal Christ-Man on the cross of creation. In Egypt there is one great temple patterned on the human figure. This is the Temple of Luxor, where Cosmic Man is figured, in both the architecture and in the ritual bas-relief designs, as in the process of being born. The Hindu architectural sutra says, 'The Universe is present in the Temple by means of proportion'.

In our time there is a convergence between the new biological science based on cybernetics and information theory, and the mystic doctrine of the Anthropocosm. The evolving universe around and within ourselves can be encountered only through the sensory instrument that we inhabit. Therefore our brains and bodies necessarily shape all our perceptions, and have themselves been shaped by the same seen and unseen energies that have shaped every perceivable thing. Body, Mind and Universe must be in a parallel, formative identity. 'Man, know thyself' was the principle of ancient science, as it is also coming to be in modern science. To quote the physicist, Robert Dicke:

> The right order of ideas may not be, 'Here is the universe, so what must man be?' but instead, 'Here is man, so what must the universe be?'
> (Quoted in C.W. Misner, K.S. Thorne, J.A. Wheeler, *Gravitation*.)

The human body contains in its proportions all of the important geometric and geodesic measures and functions. The ancient Egyptian cubit, which is a time-space commensurate measure (1/1000th of the distance that the earth rotates at the equator in one second of time), the foot, the fathom, the ancient Egyptian equivalent to the metre, all these measures are commensurate with the size or movements of the earth. The relationship of ϕ is given by the navel. In the ideal proportions of Man the arm-span in relation to the total height gives the chord-arc relation for an arc of 60°. The height of the upper body (above the hip-socket) is in relationship to the total height as the volume of a sphere is to the volume of its circumscribing cube (1:1·90983). Also the height of the upper body is to the height of the pubic arch as $\pi/3:1$ or 1·047:1. Thus the proportions of ideal man are at the centre of a circle of invariant cosmic relationships.

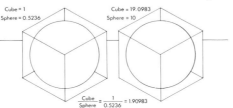

The relationship of cube to inscribed sphere.

Through an identification with the essential universal proportions expressed through this ideal human form, individual man may contemplate the link between his own physiology and universal cosmology, thereby envisaging a relationship with his own universal nature. This array of universal proportions within the body of Ideal Man becomes the basis, in many civilizations, of a canon which governs the metre for chant and poetry, the movements of dance, and the proportions of crafts, art and architecture.

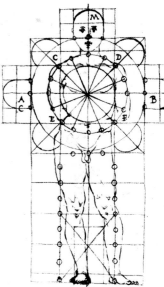

Man and Cross as the premise for the cathedral plan. In the philosophy of temple architecture the temple is to represent the image of Paradigmatic Man, the supreme archetype who emanates all of nature out of himself.

The Gothic Cathedral at Amiens, a symbolization of Universal or Cosmic Man, of whom Christ was an incarnation.

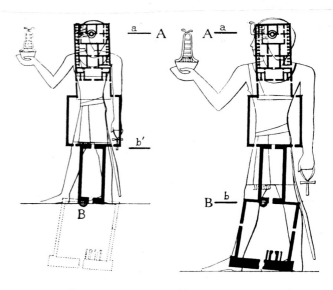

In Egypt the King was an earthly representation of the Anthropocosmic Principle, and was the motif for the construction of the Temple of Luxor. (See p. 73.)

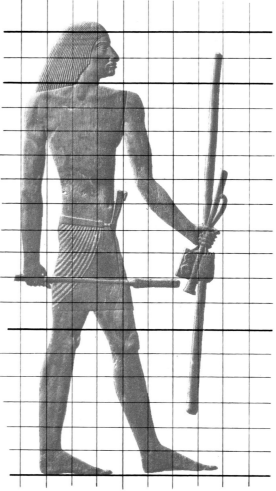

In both Renaissance and early Egyptian art we see a canon of proportions used to construct the proportions of the human body. Both these examples make use of a canon of 18 (or 9) squares from feet to brow. (See p. 86.)

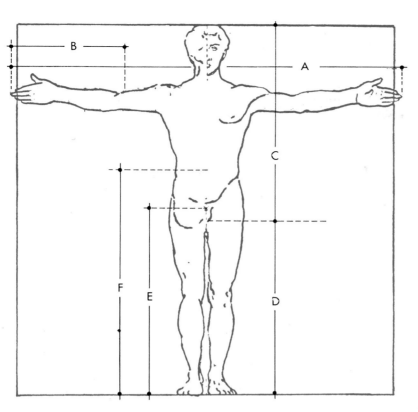

Invariant geometric and geodetic relationships are expressed in human biometrics.
  A = armspan = the fathom (four cubits)
  B = the forearm = the cubit
  C = the upper body
  D = the lower body
  E = the pubic arch
  F = the navel or ϕ division
  G = the *Hara* or $\sqrt{2}$ division = 0·586.
This vital centre is called the 'seed-pod' in
the Tantric system.

The positions of Hindu classical dance
(*Bharat Natyam*) describe geometric angular
relationships from the axis of the body's
centre of gravity just below the navel.
These positions, while defining principle
angles, are also often attributed to various
deities and are meant to convey their
characteristic powers.

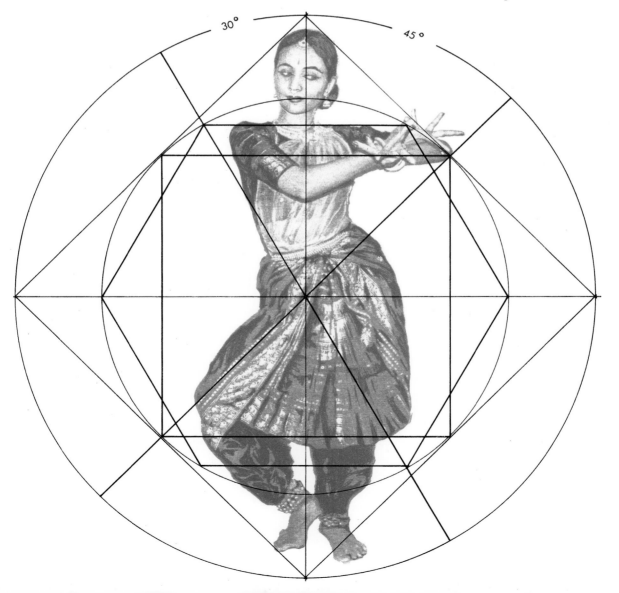

X   The Genesis of Cosmic Volumes

The perspective of *volume* offers yet another metaphor for the original and ever-continuing creative act of the materialization of Spirit and the creation of form. The very ancient creation myth coming from Heliopolis in Egypt gives an example of this mode of envisioning. Nun, the Cosmic Ocean, represents pure, undifferentiated spirit–space, without limit or form. It is prior to any extensive, any specificity, any god. It is pure potentiality. By the seed or will of the Creator, who is implicit within this Nun, the undifferentiated space is impelled to contract or coagulate itself into *volume*. Thus Atum, the creator, first creates himself or distinguishes himself from the undefinable Nun by volumizing, in order that creation might begin.

What form, then, might this first volume have? What indeed are the most essential volumetric forms? There are five volumes which are thought to be the most essential because they are the only volumes which have all edges and all interior angles equal. They are the tetrahedron, octahedron, cube, dodecahedron and the icosahedron, and are the expressions in volume of the triangle, the square and the pentagon, 3, 4, 5. All other regular volumes are only truncations of these five. These five solids are given the name 'Platonic' because it is assumed that Plato has these forms in mind in the *Timaeus*, the dialogue in which he outlines a cosmology through the metaphor of planar and solid geometry. In this dialogue, which is one of the most thoroughly 'Pythagorean' of his works, he establishes that the four basic elements of the world are earth, air, fire and water, and that these elements are each related to one of the solid figures. Tradition associates the cube with earth, the tetrahedron with fire, the octahedron with air and the icosahedron with water. Plato mentions 'a certain fifth composition' used by the creator in the making of the universe. Thus the dodecahedron came to be associated with the fifth element, aether (*prana*). Plato's fabricator of the universe created order from the primordial chaos of these elements by means of the essential forms and numbers. The ordering according to number and form on a higher plane resulted in the intended disposition of the five elements in the physical universe. The essential forms and numbers then act as the interface between the higher and lower realms. They have in themselves, and through their analogues with the elements, the power to shape the material world.

As Gordon Plummer notes in his book *The Mathematics of the Cosmic Mind*, the Hindu tradition associates the icosahedron with the Purusha. Purusha is the seed-image of Brahma, the supreme creator himself, and as such this image is the map or plan of the universe. The Purusha is analogous to the Cosmic Man, the Anthropocosm of the western esoteric tradition. The icosahedron is the obvious choice for this first form, since all the other volumes arise naturally out of it.

			Edges	Faces	Vertices	Length
Tetrahedron			6	4	4	$\sqrt{2}$
Octahedron			12	8	6	$\frac{1}{\sqrt{2}}$
Cube			12	6	8	1
Icosahedron			30	20	12	ϕ
Dodecahedron			30	12	20	$\frac{1}{\phi}$

The five regular 'Platonic' solids.

OPPOSITE The five regular polyhedra or Platonic solids were known and worked with well before Plato's time. Keith Critchlow in his book *Time Stands Still* presents convincing evidence that they were known to the Neolithic peoples of Britain at least 1000 years before Plato. This is founded on the existence of a number of spherical stones kept in the Ashmolean Museum at Oxford. Of a size one can carry in the hand, these stones were carved into the precise geometric spherical versions of the cube, tetrahedron, octahedron, icosahedron and dodecahedron, as well as some additional compound and semi-regular solids, such as the cube-octahedron and the icosidodecahedron. Critchlow says, 'What we have are objects clearly indicative of a degree of mathematical ability so far denied to Neolithic man by any archaeologist or mathematical historian'. He speculates on the possible relationship of these objects to the building of the great astronomical stone circles of the same epoch in Britain: 'The study of the heavens is, after all, a spherical activity, needing an understanding of spherical coordinates. If the Neolithic inhabitants of Scotland had constructed Maes Howe before the pyramids were built by the ancient Egyptians, why could they not be studying the laws of three-dimensional coordinates? Is it not more than a coincidence that Plato as well as Ptolemy, Kepler and Al-Kindi attributed cosmic significance to these figures?'

Concurrently, Lucie Lamy in her forthcoming book on the Egyptian system of measure gives proof of the knowledge of the five solids in the Egyptian Old Kingdom.

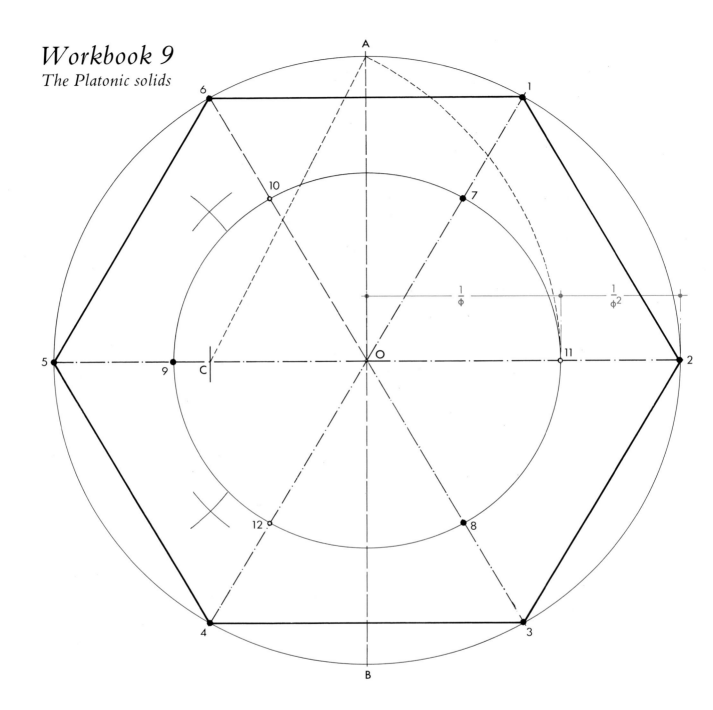

Drawing 9.1. The simultaneous generation of the
Platonic solids within the icosahedron. Draw a
circle with radius OA and inscribe hexagon (2.5) of
side $OA = 1$. Draw vertical diameter AB. Mark
each apex of the hexagon with numbers 1 to 6 and
draw in the three diagonals 1-4, 2-5, 3-6. With the
midpoint C as centre and radius CA, draw arc to
intersect radius O-2 at point 11. Line $CA = \sqrt{5}/2$
and will divide radius O-2 into the proportion $1/\phi$
and $1/\phi^2$. Draw circle with radius O-11 and at the
places where this circle intersects the radii of the
hexagon mark a point and designate with a
number 7 to 12.

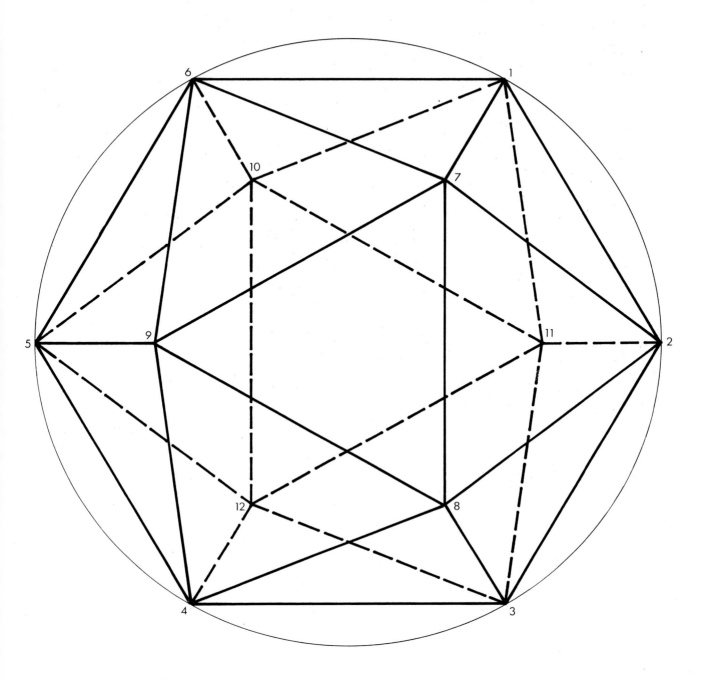

Drawing 9.2. Points 7, 8, and 9 form one of the 20 faces of the icosahedron. This face, like the other 19, is an equilateral triangle, shown here in true proportion since it is parallel to the plane of the picture. Faces 7, 8, 2; 8, 9, 4; 9, 7, 6; and 6, 7, 1; 1, 7, 2; 2, 8, 3; 3, 8, 4; 4, 9, 5; and 5, 9, 6 complete the 10 faces directly visible to the eye. Points 10, 11, 12 denote the other plane seen in true proportion. It is located directly opposite 7, 8, 9 but hidden to the eye as are the 9 other planes indicated by the dotted lines.

It can be seen that through ϕ, 'the divine seed', the icosahedron takes form.

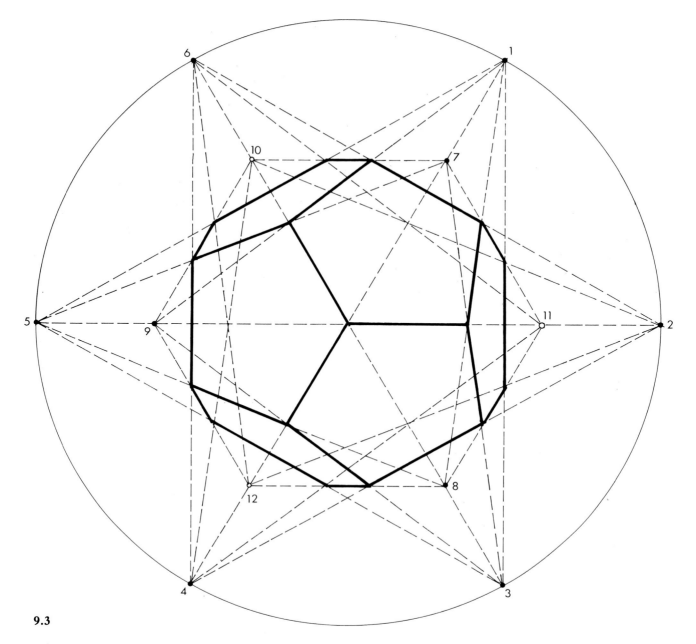

9.3

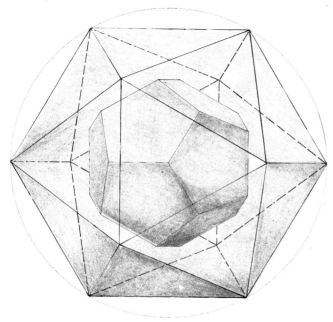

Drawing 9.3. Within a sphere of equal radius to that of the previous drawing, indicate the icosahedron by its 12 points only. Draw all connections between the 12 vertices, omitting all the diameter lines (the lines passing through the centre of the sphere). It will be seen that from each point a maximum of 5 'rays' can connect to opposite points.

For example, from point 4 draw 4-10, 4-6, 4-7, 4-2 and 4-11. (In fact 5 opposite points will define an exact pentagonal plane: 10-6-7-2-11 centred about a diameter line passing through point 4.) Repeat with points 5, 6, 1, 2, and 3, referring to Drawing 9.2 as a visual aid. From point 8 draw 'rays' 8-12, 8-5, 8-6, 8-1, and 8-11. Repeat with points 9, 7, 11, 12, and 10.

This entire set of 'rays' will intersect in groups of 3 rays at 20 points of intersection. These 20 points are the vertices which define a dodecahedron 'suspended' within the larger icosahedron. The six visible faces of the 12 are shown for visual clarity.

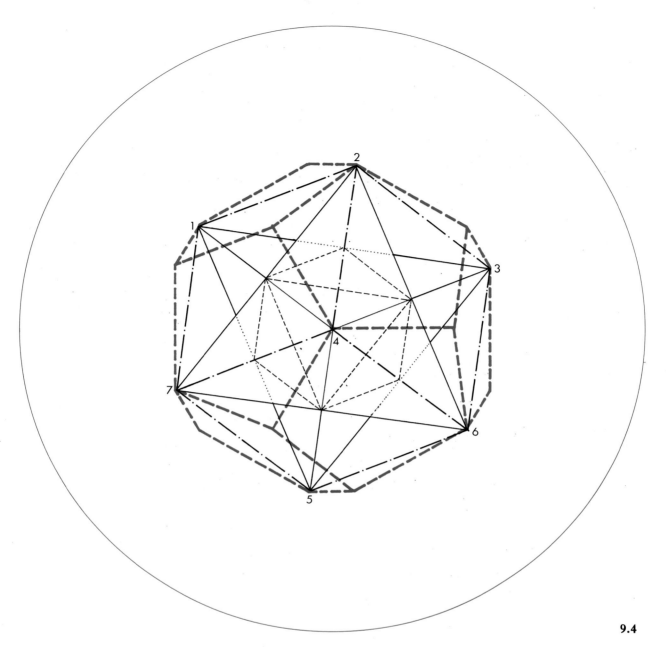

9.4

The generation of the dodecahedron arises spontaneously, a result of the natural crossing of all the internal radiants of the icosahedron. These two figures are the inverse of one another: both are composed of 30 edges, but whereas the icosahedron has 20 faces and 12 vertices, the dodecahedron has 12 faces and 20 vertices.

Drawings 9.4, 9.5. The establishment of the dodecahedron automatically gives rise to the cube defined by the 8 vertices of the dodecahedron, the edges coinciding with one diagonal of each face. Visible are the top face 1, 2, 3, 4 and two side faces 3, 4, 5, 6 and 1, 4, 5, 7. The diagonals of the faces of this cube form an interlocking or star tetrahedron. The star tetrahedron consists of two tetrahedra with points in opposite directions which are interlocked.

The volume enclosed by the two interlocking tetrahedra defines an octahedron, thus completing the composite group of regular polyhedra.

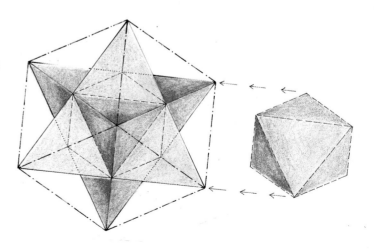

The cube is shown perfectly containing the star tetrahedron. The octahedron, like the cube, the star tetrahedron and the icosahedron, is seen in two-dimensional perspective as a hexagon. Only the dodecahedron is not contained by the outline of the hexagon.

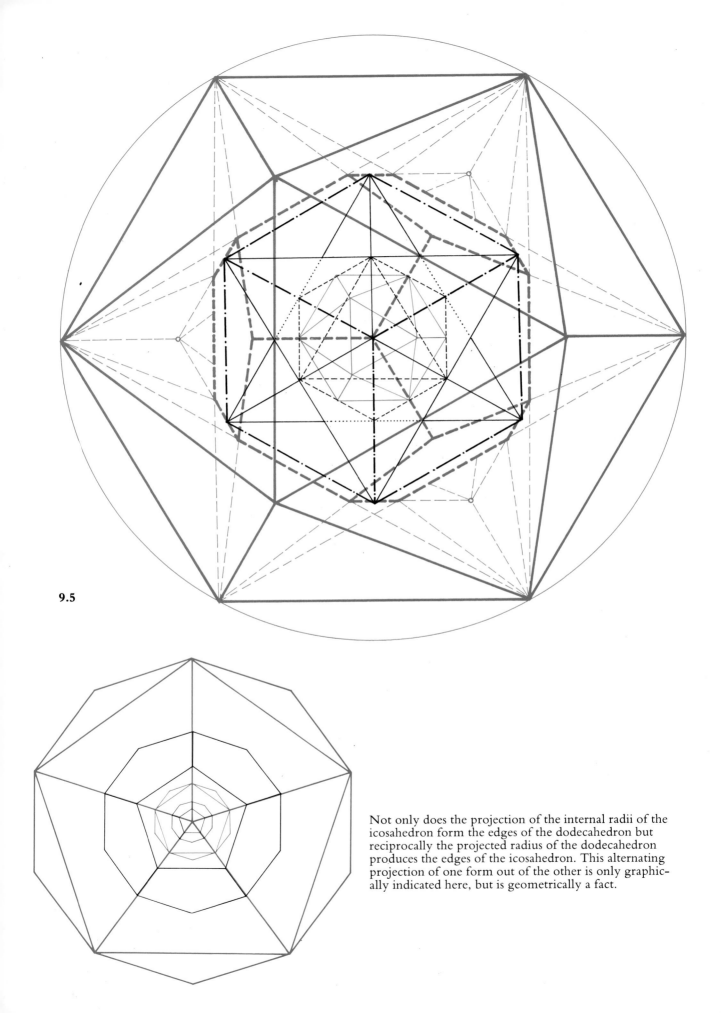

9.5

Not only does the projection of the internal radii of the icosahedron form the edges of the dodecahedron but reciprocally the projected radius of the dodecahedron produces the edges of the icosahedron. This alternating projection of one form out of the other is only graphically indicated here, but is geometrically a fact.

Let us review verbally what we have just experienced geometrically. If we connect all the internal vertices of the icosahedron by drawing three lines from each one connecting it to the ones opposite, and then from the two upper vertices draw four lines to the opposite ones, and allow these lines to converge at the centre, we will in so doing form naturally the edges of a dodecahedron (see Drawings 9.1 and 9.2). This is a generation which happens by itself through the crossing of the internal radiants of the icosahedron. Once we have established the dodecahedron we can, simply by using six of its points and the centre, form a cube. Simply by using the diagonals of the cube we are able to form the star-tetrahedron or interlocking tetrahedron. The intersections of the star-tetrahedron with the cube give us the perfect positioning to form an inscribed octahedron. Then within the octahedron, using again the lines given by the internal radiants of the icosahedron, along with the points of the octahedron, there arises a second icosahedron. We have gone through a complete cycle, through five stages, from seed to seed. This then is an infinite progression.

If the cube is given the dimension of 1, then the side of the outer icosahedron will equal ϕ and the dodecahedron will have a side length of $1/\phi$. The interlocking tetrahedron will have a side of $\sqrt{2}$. The octahedron will have a side of $1/\sqrt{2}$, and the side of the new, small, inner icosahedron will be $1/\phi^2$: a stunning constellation of harmonies. The Father (Purusha) has given birth to himself.

The single key that one needs in order to start this drawing is the method of how to find the vertices of the first icosahedron. This is given to us on the radius of a circle and is its division by ϕ.

Commentary on Workbook 9

The Hindus envisioned the Purusha as unmanifest and untouched by creation just as in the drawing the icosahedron is untouched by the other forms. The dodecahedron was then seen to be Prakriti, the feminine power of creation and manifestation, the Universal Mother, the quintessence of the natural universe. This dodecahedron touches all the forms of creation within her silent, observing partner. The interlocking tetrahedron is then seen as the yin and the yang, for the tetrahedron is a volume of threeness and is therefore a primary symbol of a function accompanied by its reciprocal. The result of this harmonic interaction of opposites gives the cube, symbolic of material existence, the four states of matter, earth, air, fire, water. Both the cube and the interlocking tetrahedron touch the dodecahedron. At the heart of this tetrahedron is the octahedron, and as the cube is a formation of its extremities, the octahedron symbolizes the crystallization, the static perfection of matter. It is the diamond, the heart of the cosmic solid, the transformed and clarified lens of light, the double pyramid. The outer progression, extending into vaster and vaster realms, demarcates the same progression, the same genesis: icosahedron, the Purusha, generating the dodecahedron, the Prakriti, and within Prakriti the full play of manifested existence. The whole coagulation is begun by the secret seed which contracts the circle, the infinite, undifferentiated spirit, into the icosahedron. The seed is phi, the fire of spirit.

The transcendent principles, the icosahedron and the dodecahedron, Purusha and Prakriti, the primal duality, each have phi proportions. But when we arrive at the level of the natural world of oppositional dualities, yin and yang, and the cube of matter and its crystallization in the octahedron, it is the square root of 2 which is active. The square root of 2 is the way in which ϕ acts in Nature. And from the octahedron, the purified state of matter, its crystallization into the mineral gem, is reborn the icosahedron with its phi dimension, $1/\phi^2$. This proportion, $1/\phi^2 = 0.382$ . . . is the geometric function associated with Christ (see p. 63). Being a square, it represents a manifested form, the Son; and being the side of the inner icosahedron it is the incarnation or exact image of the initial, generating icosahedron, the Father, Purusha, Anthropocosm.

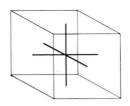

Two cubes of halite

The mineral world expresses pure volumetric geometry with the greatest clarity, but it is important to remember that these solids do not exist in nature. In their perfect form they exist only on a metaphysical plane, as pure, creative ideation, and can be represented, for the mind to grasp, only through geometry.

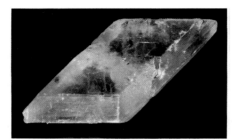

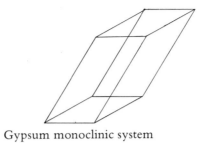

Gypsum monoclinic system

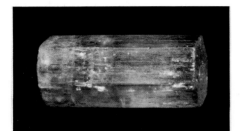

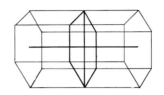

Beryl hexagonal system

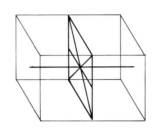

Quartz trigonal system

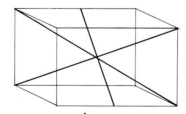

Idocrase tetragonal system

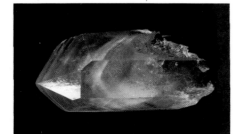

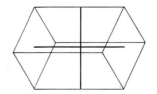

Chlorite in quartz

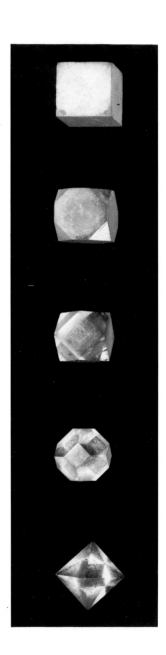

Purusha and Prakriti are the eternal creative dichotomy in Hindu mythology. Purusha is the anthropocosmic, paradigmatic Man or Seed that projects Prakriti, the eternally enchanting Feminine, in order that her womb may give birth to his own embodiment in the world of form.

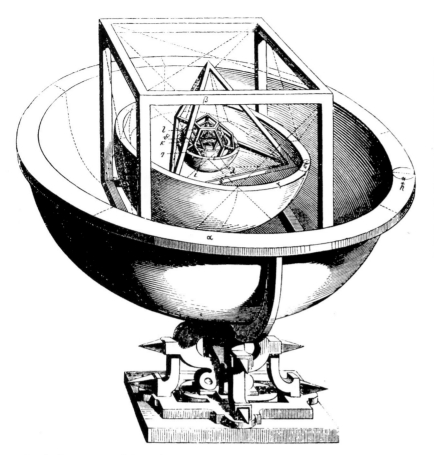

OPPOSITE In this demonstration the regular polyhedra are determined by nine concentric circles whose pattern gives all the necessary information for the construction of these forms. Each volume is in a simple harmonic relationship to the others, and it is the same transcendental functions, $\sqrt{2}$, $\sqrt{\phi}$ and ϕ that make up these patterns of relationships. As in the previous drawing, all the volumes emerge simultaneously. But in this case if one of the concentric circles is removed then the pattern cannot yield the remaining volumes. This is an image of the great Buddhist idea of the co-dependent origination of the archetypal principles of creation.

Kepler's version of the solar system was as one Platonic solid within another, the radii of the intervening concentric spheres relating to the orbits of the planets.

These symbolic volume-forms symbolically re-enact our cosmic history, and perfectly represent the great movements whose meanings they convey. The play is that of the constant interchange between the icosahedron as the male Purusha and the dodecahedron as the feminine Prakriti. The icosahedron is a structure of 12 vertices and 20 faces. It is a structure of triangles, three being the dynamic 'male' number. The androgynous dodecahedron as giver of life has 12 faces and 20 vertices and is a structure of five, the number of life (3 male + 2 female). The star born within its pentagon is the configuration of Cosmic Man, the perfector of life, the Golden Proportion.

These same five regular volumes are classically drawn in such a way that they are contained within nine concentric circles, with each solid touching the sphere which circumscribes the next solid within it. This design will yield many important relationships and is derived from the discipline called *corpo transparente*, of contemplating the shapes, constructed of transparent material, placed one within the other. This instruction was given to many of the great men of the Renaissance, including Leonardo, Brunelleschi and Giorgi, by Fra Luca Pacioli.

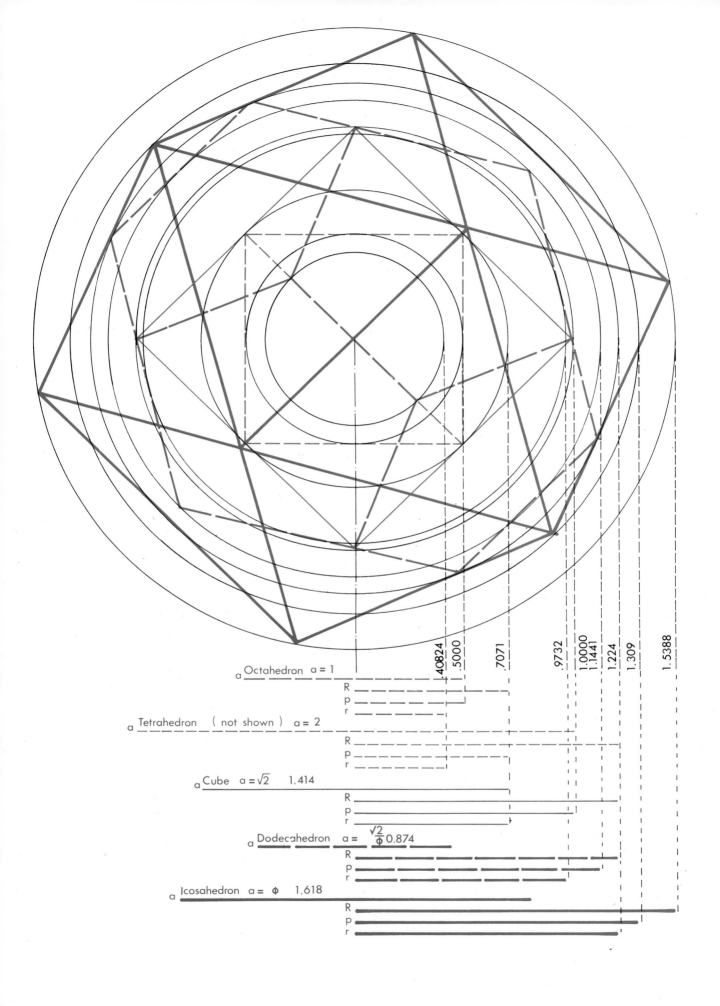

.40824
.5000
.7071
.9732
1.0000
1.1441
1.224
.309
1.5388

ₐ Octahedron a = 1
R
p
r

ₐ Tetrahedron (not shown) a = 2
R
p
r

ₐ Cube a = √2 1.414
R
p
r

ₐ Dodecahedron a = $\frac{\sqrt{2}}{\Phi}$ 0.874
R
p
r

ₐ Icosahedron a = Φ 1.618
R
p
r

Fra Luca Pacioli, the great Renaissance teacher of sacred geometry. The student's concentration on the transparent solids was a discipline to assist in seeing the metaphysical realities beneath all appearance.

There is speculation that in Hindu metaphysics each one of the bodies was the symbol of one of the invisible, subtle envelopes which were believed to surround and interact with the physical body of man. The tradition associates

Individualizing
{
  the small central icosahedron with the ultimate Perfection of the Body in its physical manifestation;

  the octahedron with the physical or Food Body (seat of the instinctual mind);

  the tetrahedron with the pranic or Energetic Body (seat of the intuitive mental faculty);
}

Transpersonal
{
  the cube with the Mind-Body of 'pure reason';

  the dodecahedron with the Knowledge Body (seat of innate knowledge by identity);

  the icosahedron with the Bliss Body (that of meditative union).
}

In conclusion we may ask how the practice of Sacred Geometry helps us confront the profound questions of existence: What is the nature of Spirit? What is the nature of Mind? What is the nature of Body?

My individual practice of Geometry gives this reply: The Body is the most dense expression of Mind, and Mind is all the subtle extensions of Body; and underlying this entire world, from the most dense to the most subtle, there is one substance. This substance is Spirit which has become entranced by the beauty of geometrizing.

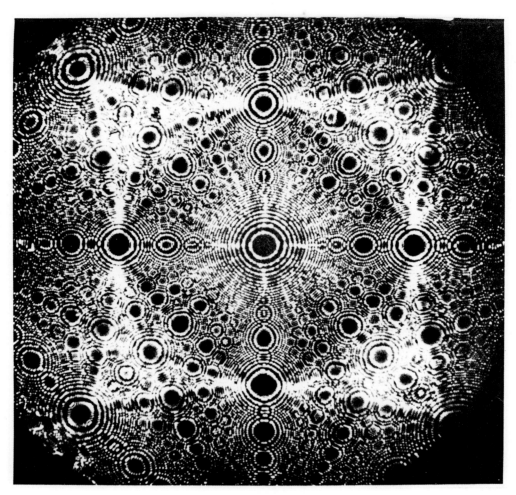

These refraction photos are the closest visualization that science can give with respect to the nature of atomic substance, which appears to be patterns of geometrized light-energy.

Bibliography

Acts of John, The, Apocrypha of the New Testament.

Aurobindo, Sri, *The Life Divine*, Centenary Edition, vols 18, 19, Pondicherry, India, Sri Aurobindo Ashram Trust, 1970.

Beckmann, Petr, *A History of Pi*, New York, St Martin's Press, 1971.

Boyer, Carl B., 'Zero, the Symbol, the Concept, the Number', *National Mathematics Magazine*.

Brunés, Tons, *The Secrets of Ancient Geometry – and its Use*, 2 vols, Copenhagen, Rhodos, 1967.

Charpentier, Louis, *The Mysteries of Chartres Cathedral* (trs. Sir Ronald Fraser), London, Research into Lost Knowledge Organization, 1972.

Colman, Samuel, N. A., *Nature's Harmonic Unity*, New York, Putnam's, 1912.

Critchlow, Keith, *Islamic Patterns*, London, Thames & Hudson, New York, Schocken, 1976.

——, *Order in Space*, London, Thames & Hudson, 1969, New York, Viking, 1970.

——, *Time Stands Still*, London, Gordon Fraser, 1979, Forest Grove (Ore.), International Scholarly Book Service, 1980.

Daniélou, Alain, *Traité de musicologie comparée*, Paris, 1959.

——, *Hindu Polytheism*, Bollingen Series LXXIII, New York, Pantheon Books, 1964.

Danzig, Tobias, *Number, the Language of Science*, New York, The Free Press, 1967.

De Nicolás, Antonio, *Avatara*, New York, Nicolas Hays, 1976.

Fabre d'Olivet, Antoine, *La Musique expliquée comme science et comme art*, Collection Delphica, Lausanne, Age d'Homme, 1972.

Fournier des Corats, D., *La Proportion égyptienne et les rapports de divine harmonie*, Paris, Editions Vega, 1957.

Ghyka, Matila, *Esthétique des proportions dans la nature et dans les arts*, Paris, Gallimard, 1933.

——, *The Geometry of Art and Life*. New York, Dover, 1977.

——, *Le Nombre d'or*, 3 vols, Paris, Gallimard, 1931.

Gillings, Richard J., *Mathematics in the Time of the Pharaohs*, Cambridge (Mass.), London, MIT Press, 1972.

Guénon, René, *Les Principes du calcul infinitésimal*, Paris, Gallimard, 1946.

Hambidge, Jay, *The Elements of Dynamic Symmetry*, New York, Dover, 1967.

Heninger, S. K., Jr., *Touches of Sweet Harmony*, San Marino, California, Huntington Library, 1974.

Jenny, Hans, *Cymatics I and II*, Basel, Basilius Press, 1974.

Kramrisch, Stella, *The Hindu Temple*, 2 vols, Delhi, Motilal Banarsidass Press, 1976.

Levarie, Siegmund, and Ernst Levy, *Tone: A Study in Musical Acoustics*, Kent State University Press, 1968.

McClain, Ernest, *The Myth of Invariance*, New York, Nicolas Hays, 1976; Boulder (Col.), London, Shambhala, 1978.

——, *The Pythagorean Plato: Prelude to the Song Itself*, New York, Nicolas Hays, 1978.

Menninger, Karl, *Number Words and Number Symbols*, Cambridge (Mass), MIT Press, 1970, London, MIT Press, 1977.

Michel, Paul Henri, *De Pythagore à Euclide: contribution à l'histoire des mathématiques préeuclidiennes*, Paris, Belles Lettres, 1950.

Michell, John, *City of Revelation*, London, Garnstone Press, 1972, New York, Ballantine, 1977.

Néroman, D., *Les Leçons de Platon*, Paris, Niclaus, 1943.

Nicomachus of Gerasa, *Introduction to Arithmetic* (trs. Martin Luther D'Ooge), New York, Macmillan, 1926; in Euclid, *The Thirteen Books of Euclid's Elements*, Cambridge, Cambridge University Press, 1926.

—— (Nicomachus Gerasenus), *Manuel d'harmonique et autres textes rélatifs à la musique* (trs. C.-E. Ruelle), Paris, Baur, 1881.

Pauling, Linus and Hayward, Roger, *The Architecture of Molecules*, San Francisco, London, W. H. Freeman, 1964.

Peet, Eric, *The Rhind Mathematical Papyrus*, Liverpool, Liverpool University Press, London, Hodder and Stoughton, 1923; Reston (Va.), National Council of Teachers of Mathematics, 1979.

Plato, *Timaeus* (trs. Thomas Taylor), Minneapolis, Wizard's Bookshelf, 1975.

Purce, Jill, *The Mystic Spiral*, London, Thames & Hudson, 1974, New York, Thames & Hudson, 1980.

Schwaller de Lubicz, R. A., *Le Miracle égyptien*, Paris, Flammarion, 1963.

——, *Le Roi de la théocratie pharaonique*, Paris, Flammarion, 1961 (Forthcoming in English from Inner Traditions International, New York.)

——, *Symbol and the Symbolic*, (trs. R. and D. Lawlor), Brookline (Mass.), Autumn Press, 1978.

——, *Le Temple de l'homme*, Paris, Caractères, 1957, trs. R. and D. Lawlor as *The Temple in Man*, Brookline (Mass.), Autumn Press, 1977.

Schwenk, Theodore, *Sensitive Chaos*, London, Rudolf Steiner Press, 1965, New York, Schocken, 1978.

Smith, D. E., *History of Mathematics*, 2 vols, New York, Dover, 1958.

Suarès, Carlo, *The Cipher of Genesis* (trs. from the French), London, Stuart & Watkins 1970, New York, Bantam Books, 1973.

Theon of Smyrna, *The Mathematics Useful for Understanding Plato* (trs. from the Greek/French edn of J. Dupuis by R. and D. Lawlor), San Diego (Cal.), Wizard's Bookshelf, 1979.

Thompson, D'Arcy, *On Growth and Form*, Cambridge University Press, 1971.

Toben, Bob, *Space, Time and Beyond*, New York, Dutton, 1975.

VandenBroeck, André, *Philosophical Geometry*, South Otselic, New York, Sadhana Press, 1972.

Vitruvius, *Ten Books on Architecture*, New York, Dover, 1960.

Warusfel, André, *Les Nombres et leurs mystères*, Paris, Seuil, 1961.

Young, Arthur, *The Geometry of Meaning*, New York, Delacorte Press, 1976, London, Wildwood House, 1977.

Sources of Illustrations

4 Photo Lane Eastman Kodak Co.

4 Silkscreen, Auroville, S. India, 20th c.

5 Sole X-ray. Photo Dr Wolf Strache.

5 West rose with overlay, Chartres Cathedral, France, c. 1216. Photo Painton Cowen.

7 G. Riesch, *Margarita philosophica*, Basle 1583.

7 G. Riesch, *Margarita philosophica*, Freiburg 1503.

7 F. Gaffurio, *Theorica musica*, Milan 1492.

8 Photo Science Museum, London.

9 Gouache on cloth, Nepal, c. 1700. John Dugger & David Medalla, London.

10 Silvacane abbey, France, 12th c. Photo F. Walch, Paris.

11 'The Creator', Bible Moralisée, France, c. 1250. Bodleian Library, Oxford.

13 Brush painting by Sengai, Japan, c. 1830. Mitsu Art Gallery, Tokyo. Photo Arts Council of Great Britain.

14 Science Museum, London.

15 'The Yogi and his Symbols', ink and gouache on paper, Rajasthan, c. 18th c. Ajit Mookerjee.

16 Dome of Capilla del Condestable, Burgos cathedral, Spain, 1482–94. Photo Mas.

16 Mandala tanka, Tibet c. 1800. John Dugger & David Medalla, London.

17 'Wheel of Law', bronze statue of Yakushi (detail), Yakushi Temple, Japan, 7th c. Photo Toshio Watanabe.

17 'Johann Neudorfer and son' (detail), painting by Nicolaas Neufchatel, 1561. Alte Pinakothek, Munich. Photo Blauel, Munich.

21 Drawing after Hans Kayser, *Lehrbuch der Harmonik*, Basle 1950.

22 Gouache and silver on paper, Rajasthan, 18th c. Ajit Mookerjee. Photo Jeff Teasdale.

22 Egyptian mouth symbol, detail from wood relief depicting Maat, goddess of Truth, from tomb of Sethos I, Egypt, 19th dynasty. Archaeological Museum, Florence. Photo Alinari/Anderson.

22 Vibrating string. Photo Science Museum, London.

24 Man as the microcosm of the four elements, astronomical ms., Prüfening, Bavaria, late 12th c.

Österreichische Nationalbibliothek, Vienna.

29 Drawing with geometric analysis of the Parthenon, after Tons Brunés, *Secrets of Ancient Geometry*, 1967.

29 Tile decoration from Bedi Palace, Marrakesh, Morocco. Photo Roland Michaud.

29 Honey bee with geometric overlay. Drawing after Samuel Colman, *Nature's Harmonic Unity*, 1912.

30 Photo Ewing Galloway (Aerofilms).

34 Marble relief, Saint-Sernin, Toulouse, France, late 11th c. Photo Jean Roubier.

35 Diagram of Chapel of St Mary, Glastonbury. Drawing by Keith Critchlow.

38 Lindisfarne Gospels, English, c. 700. British Library, London.

43 Three variations of diatoms. British Museum, London (Natural History).

43 Four Renaissance ground plans: Brunelleschi, conjectural reconstruction and plan of S. Maria degli Angeli, Florence; Serlio, from the *Fifth Book of Architecture*; Barozzi da Vignola, plan of Palazza Farnese, Caprarola; Bramante, plan of St Peter's, Rome.

53 S. Maria Novella, Florence. Photo Martin Hürlimann.

Hermes (Medusa), Roman marble after Greek original, 1st c. BC. Glyptothek, Munich.

54 Drawing of mummy of Shishou, E. wall of chapel in tomb of Petosiris, Egypt, c. 300 BC.

57 Photo F. Paturi.

58 'Asclepias Speciosa', from K. Blossfeldt and E. Weber, *Art Forms in Nature*, 1932.

59 'Vitruvian Man', drawing by Leonardo da Vinci, c. 1490. Academy, Venice. Photo Soprintendenza alle Gallerie di Venezia.

59 Canon figure, drawing by Albrecht Dürer.

60 Central hall, Abydos, looking west. From H. Frankfort, *The Cenotaph of Seti I at Abydos*, vol II, 1933.

60 Tomb sarcophagus of Osiris, from Abydos. Archaeological Museum, Marseilles.

63 National Gallery, London.

64 'Holy Trinity', Lothian Bible, c.

1220. Pierpont Morgan Library, New York (Ms 791, f. 4v.).

66 Vishnavata Temple, Khajuraho, India, 11th c. Photo Ellen Smart.

66 Ms. illustration on temple building methods, from recto and verso of a palm leaf, India.

66 Ground plan of Vaikunthaperumal Temple, Kanchipuram, India, 8th c.

71 'Briza Maxima', enlarged x15, from K. Blossfeldt and E. Weber, *Art Forms in Nature*, 1932.

71 (margin) Drawing after Carl Sagan, *The Dragons of Eden*, 1977.

71 (bottom) Drawings and texts, reproduced by kind permission of Frank J. Swetz, from *Was Pythagoras Chinese?*, F.J. Swetz and T.I. Kao, Penn State Press, 1977.

72 Osiris enthroned. Painting by Lucie Lamy, 20th c.

73 Drawing after R. A. Schwaller de Lubicz, *Le Temple de l'homme*, III, 1957.

76 Photo and drawing of carved stone capital from cathedral of Le Puy, France.

79 Photo Al Araby Magazine.

83 F. Giorgi, *De harmonia mundi*, 1525.

86 G. Valla, *De expetendis et fugiendis rebus opus*, 1519.

86 A. Dürer, *Vier Bucher von menschlicher Proportion*, 1528.

87 (above left) Photo Hans P. Widmer.

87 (below left) Ajit Mookerjee.

87 (right) Photo J. C. Stuten.

89 Carved limestone relief from tomb of Paatenemheb, Saqqara, Egypt, c. 1330 BC. Rijksmuseum van Oudheden, Leiden.

91 'Man the Procreator', from V. Scamozzi, *L'idea dell' architettura universale*, 1615.

92 Arawa wheat apical meristem at the late vegetative stage and Saltbush flower anther. Photos taken through scanning electron microscope, from J. Troughton and L. A. Donaldson, *Probing Plant Structure*, 1972.

92 Ms. illustration from ancient manual of architecture, India.

93 Study of proportions relating a basilica to the human body. Pen drawing by F. di Giorgio (1439–1501/2), Italy. Biblioteca Nazionale, Florence (Codex Magliabechiano).

93 Amiens cathedral, engraving, 19th c.

Photo Conway Library, Courtauld Institute of Art, London.

94 Temple of Luxor, Egypt. Photo Hirmer.

94 Relation of proportions of ground plan of Temple of Luxor, Egypt, to the figure of a man. Drawing from R. A. Schwaller de Lubicz, *Le Temple de l'homme*, 1957.

94 'St Christopher and the Christ Child', drawing by D. Bramante (1446–1516), Italy. Statens Museum for Kunst, Copenhagen.

94 The royal scribe Hesire, wood relief from tomb of Hesire, Saqqara, Egypt, 3rd Dynasty. Egyptian Museum, Cairo.

96 Complete set of Scottish neolithic 'Platonic Solids'. Photo Graham Challifour.

104 Crystal systems. Photos Geological Museum, London.

105 Stone sculpture from cave temple, Badami Village, India, *c.* 6th c. Photo R. Lannoy.

106 'Harmony of the Universe', from Kepler, *Mysterium Cosmographicum*, 1621.

108 'Fra Luca Pacioli and Pupil', painting by J. de Barbari (1440/50–1516). Capodimonte Museum. Naples. Photo Scala.

109 'Dancing Atoms'. Photo Dr Erwin Müller, Pennsylvania State University.